How To Edit A SCIENTIFIC JOURNAL

The Professional Editing and Publishing Series

This volume is one of a series published by ISI Press®. The Professional Editing and Publishing Series provides timely, practical information to help publishers and editors of professional, scholarly and scientific publications.

Books published in this series:

How to Edit a Scientific Journal
  by CLAUDE T. BISHOP

Copywriter's Handbook: A Practical Guide for Advertising and Promotion of Specialized and Scholarly Books and Journals
  by NAT G. BODIAN

How To Edit A SCIENTIFIC JOURNAL

Claude T. Bishop

iSi PRESS®

Philadelphia

Published by

iSi PRESS® A Subsidiary of the
Institute for Scientific Information®
3501 Market St., Philadelphia, PA 19104 U.S.A.

© 1984 ISI Press

Library of Congress Cataloging in Publication Data

Bishop, Claude T., 1925–
How to edit a scientific journal.

Bibliography: p.
Includes index.
1. Technical editing. I. Title.
T11.4.B57 1984   808.02   84–12875
ISBN 0–89495–033–9
ISBN 0–89495–034–7 (pbk.)

Printed in the United States of America
90 89 88 87 86 85 84 83   8 7 6 5 4 3 2 1

THE EDITOR'S PASSPORT

The Editor stood 'fore the Heavenly Gate,
His features pinched and cold.
He bowed before the Man of Fate,
Seeking admission to the fold.
"What have you done," St. Peter asked,
"To gain admission here?"
"I was the Journal's editor, Sir,
For many a weary year."
The Pearly Gates swung open wide
as Peter pressed the bell.
"Come in and choose your harp," he cried;
"You've had your share of hell!"

Anonymous: Published in an
editorial in *The Journal of the
Irish Medical Association*, Vol.
42, No. 247, Jan. 1958, pp. 31–32

Contents

Preface

In recent years there have been frequent expressions of concern about erosion of quality in the scientific literature as a consequence of the compulsion to "publish or perish." General statements have been made that journals should have higher standards, but there have been few specific suggestions about how to achieve that goal.

The maintenance of standards in scientific journals rests squarely in the hands of their editors, who decide what will be published. There are many journal editors, even more associate editors, and many more of both to come because editorial appointments rotate with reasonable frequency. In fact, all scientists who publish actively can confidently expect to be offered appointments as editors, associate editors, or members of editorial boards, at some stage in their careers. However, the majority of scientists who accept editorial appointments receive precious little guidance. The lucky ones may receive some advice, and inherit some forms and files, from their predecessors, but most will end up learning the business by hard and sometimes bitter experience.

This book is an attempt to fill that information gap and is based on the premise that editors of scientific journals *must be active scientists* whose main activity is research. As such, they have neither the skills nor the time to be involved directly in copy editing, printing, financial management, distribution, or marketing. These latter activities are considered here as *publishers' functions* in which scientific editors (i.e., scientists) can be involved only indirectly, on a need-to-know basis. The book is therefore concerned primarily with the review process in the production of the scientific literature and addresses some questions that do not seem to have been given the attention they deserve: Who should be an editor? Why be an editor? What are the proper functions of an editor? How should referees be selected? How can good referees be found and retained? What is the role of associate editors? What are the needs of an editorial office? What ethical matters are important for authors, editors, and referees? To what extent should editors be involved in publishers' functions? The answers to these and other questions are derived from material and experience accumulated during my involvement in the editorial process over some 17 years, first as a journal editor and then as editor-in-chief of a series of 12 journals.

I hope that this book will provide assistance and encouragement to all scientific editors and associate editors, old and new, and to the publishers who select them. It should also be of interest to authors, editors' assistants, copy editors, and printers; in short, to anyone who wants to know what goes on, or should go on, in the editorial office of a scientific journal.

I am deeply indebted to a number of people who assisted in various ways in the preparation of this book. Dr. J. A. Morrison, McMaster University, who as a former editor-in-chief first introduced me to the editorial process, was most generous in furnishing background material and was also kind enough to review an early draft of the manuscript. Mrs. Barbara Drew and Dr. Gabrielle Adams, manager and former manager of the Research Journals published by the National Research Council of Canada, provided much useful information and advice. The American Chemical Society (C. R. Bertsch), the American Geophysical Union (A. F. Spilhaus, Jr.), and the American Institute of Physics (J. T. Scott) kindly supplied samples of the forms used in their editorial offices and granted permission for their reproduction here. Miss Evelyn Monson transcribed the author's scrawl into flawless typescript and suffered through several revisions. Finally, Bob Day's enthusiastic encouragement provided a powerful stimulus without which this book would not have been written.

Chapter 1

The Literature of Science

Origins

The literature of science began in 1665 with the near-simultaneous publication of the *Philosophical Transactions of the Royal Society of London*, in England, and the *Journal des Sçavans*, in France. Prior to the initiation of those journals, scientists communicated their results by letter and by private publication of books, treatises, or pamphlets which were circulated among colleagues to pass along interesting observations and ideas. Science, more than any other human activity, depends vitally upon the printed word as a record of results that can be referred to and used for further development. Thus, when the secretaries of the Royal Society of London and the Académie des Sciences began to collect and circulate letters and reports of meetings in a systematic way, it is not surprising that their efforts were greeted enthusiastically by the scientists. The "journals" were quickly adopted as the standard means of communicating new scientific discoveries. The success of these first endeavors in scientific publication inspired other learned societies to initiate their own journals, and so it all began.

Development

Studies of the growth of science (e.g., see Price[22]) have shown a doubling time of 10 years starting from about the year 1700. Individual journals were hard pressed to keep pace with this rate of growth. As a consequence, some existing journals were split and new journals were started, each of them dealing with finer and finer subdivisions of a field of knowledge. The trend to specialized and subspecialized journals has continued unabated with dire predictions that we shall soon see titles such as "The Journal of Pine Tree Needles" or even "The Journal of Lonesome

1

Pine Tree Needles." The reader who is interested in a more scholarly and detailed treatment of the development of scientific journals is directed to two historical accounts.[17,19]

Concern over the proliferation of scientific publications is by no means new. Witness the following excerpt from an editorial published in the first volume of the *Annual Report of the Progress of Chemistry:*

> The great and daily increasing number of researches appearing every year in the different departments of Chemistry and the Allied Sciences, renders it difficult for individuals to obtain, by actual inspection of the original sources, a complete survey of their progress.
>
> The study of one, or even of several journals, does not suffice for this purpose, the communications of various investigators being distributed over a large number of periodicals, and many papers, especially interesting to the Chemist, being actually buried in publications chiefly devoted to other subjects.

Apart from some minor peculiarities of language, this editorial could have been published today, most likely as a letter to the editor of *Science* or *Nature*. The fact that it was published in 1848 may put the topic in perspective.

Parkinson[21] has addressed the topic of proliferation of scientific journals in a less serious vein:

> Why, to begin with, should they multiply? Because each must fall, sooner or later, into the clutches of a professor (A) more fanatically jealous than the average. Under his editorship no article is accepted with which he does not agree and no book kindly reviewed other than those written by his own former pupils. The rival professor (B) whose articles have been most consistently excluded will then, and inevitably, start another journal; one edited at first on more liberal principles. B will accept articles from all who are not actual and known adherents of A. He eventually draws the line, however, at contributions from C, whose works are confused, long, and original only in their grammar and punctuation. But C realizes by now where his remedy lies. He becomes the founder of a new and less hidebound periodical; one more open at first to new and confused ideas. There is difficulty, in the end, however, over articles submitted by D, who cannot even spell. But D is not to be denied access to the misprinted page. He hesitates, to be sure, before adding one more journal to the library shelves; but not for long. His duty is clear and he does not shirk it.

As with most of Parkinson's concepts, there is just enough truth in the humorous exaggerations to make us feel a bit uncomfortable.

Current Status

How Many Journals?

All scientists would agree that there are too many journals being published today. The total number is unknown; estimates range from a high of 90,000 to a low of 30,000. The difficulty in arriving at a valid estimate is one of definition. What constitutes a scientific journal? There is probably no generally acceptable answer to this question, but most active scientists would argue that the number of *significant* scientific journals is much less than 30,000. What then is a significant journal? A possible definition is that a significant journal is one that participates effectively in the transfer of scientific information; someone, somewhere, reads an article in the journal, uses the information, and cites it. *Journal Citation Reports*, published annually by the Institute for Scientific Information, Philadelphia, lists the journals that were cited during a yearly period in the journals included in the *Science Citation Index* data base. The total number listed is around 4,000. Even if one wants to argue with the journal coverage of the *Science Citation Index*, it would not be out by more than a factor of 2. The number of significant scientific journals is therefore no more than 8,000. It could also be argued that a journal that is cited only once, or even twice, is not participating very effectively in the transfer of information, but the line has to be drawn somewhere.

Suddenly, by applying a significance factor, the scientific literature is reduced from an estimated 30,000 to 90,000 journals to a much more manageable 8,000 or less. Of course, individual scientists practice further discrimination by limiting their reading to a small group of journals in which they, and the other members of their specialty, publish the majority of their papers. Does this mean that the information explosion is a myth? Not at all, but the explosion is only real for those who, perhaps mistakenly, try to cover the whole scene.

Information Retrieval

Of course, for journals to serve as the record of science, the information that is in them must be retrievable. The early journals developed indexes, and then came abstract journals that published short summaries of all papers in a given field, indexed by subject, author, title, and, more recently, by key words. Perhaps the most useful development has been the increasing numbers of review journals with titles such as "Annual Review of . . . ," "Annual Reports of . . . ," "Advances in . . . ," "Current Topics in. . . ." These journals publish status reports of specific areas of research with all pertinent references. They are extremely useful in keeping scientists up to date or in introducing them to a new field. Finally, there are the information systems that utilize the computer to search large data bases by author, title, or key words. The enormous capacity of the computer has encouraged the developers of such systems to strive for comprehensive coverage, and many data bases will include items from unrefereed reports, abstracts of meetings, conference proceedings, and the like. The term "GIGO" (garbage in, garbage out) applies; only a small percentage of the items turned up in a search of such "comprehensive" data bases are likely to be useful. To be sure, a comprehensive search can provide some assurance that a key reference has not been overlooked. Also, scientists can save many a trip to the library with a carefully constructed key word profile for retrieving items from the current literature. However, computerized data bases cannot be browsed and tend to be overloaded because the compulsion for comprehensive coverage leaves no room for qualitative selectivity. The thought processes of scientists are complex and diverse, but browsing and qualitative judgments are certainly part of them and are not amenable to computerization. A new and exciting idea may be triggered, not by the article that is turned up by a computer search, but by the one that is printed next to it in the same issue of the journal. Scientists also exercise qualitative judgments in their reading and will not waste time looking at journals that they know by experience are collections of trivia or contain nothing of interest in their field.

The Primary Research Journal

Collectively, the journals, reviews, abstracts, indexes, and retrieval systems constitute the literature of science. There is little need to emphasize that the heart of this literature is the primary research journal upon which all the other parts depend and from which they are derived. Primary research journals are those that publish the first reports of original research. The key words in this definition are "first," which means that the work has not been published before, and "orginal," which means that the research reported is a new contribution to knowledge. From this definition it is clear that primary journals are in fact an integral part of the scientific process. Science usually progresses by small steps, with new studies being stimulated by, and building upon, the results of earlier research. Scientists recognized early that progress would be faster if there was a way to communicate and record the results of new investigations, to make them available to all scientists. The urge to communicate quickly led quite naturally to what we know as the scientific "paper," that is, a brief description of a definitive piece of research that contributes specific new knowledge to its field. The publication of these short reports permitted the rapid circulation of ideas and information throughout the scientific community where they could be tested, verified, and debated—the process of becoming part of "the consensus of science."[27] The best indication of the efficiency of this process is that the general format of the scientific paper has not changed in over three centuries. An entertaining example is a paper published in the *Philosophical Transactions of the Royal Society*, vol. 60, pp. 54–64, 1770; it is in the form of a letter from Daines Barrington, F.R.S., to Mathew Maty, M.D., Sec. R.S. The paper is entitled "Account of a Very Remarkable Young Musician" and provides proof that Mozart was indeed a child prodigy. The first similarity to a modern paper is the prominent notation that the manuscript was "received November 28, 1769." Then comes a very brief introduction:

> If I was to send you a well attested account of a boy who measured seven feet in height, when he was not more than eight years of age, it might be considered as not undeserving the notice of the Royal Society.
> The instance which I now desire you will communicate to that learned body, of as early an exertion of most extraordinary musical talents, seems perhaps equally to claim their attention.

Admittedly, this brief statement sets up a straw man in the first paragraph. However, it states clearly the subject of the paper and why it should be of interest.

The paper then continues with proof of Mozart's date of birth and a description of tests (experimental) that the author performed with the young boy.

> I carried to him a manuscript duet, which was composed by an English gentleman. . . . My intention in carrying with me this manuscript composition, was to have an irrefragable proof of his abilities, as a player at sight, it being absolutely impossible that he could have ever seen the music before.
> The score was no sooner put upon his desk, than he began to play the symphony in a most masterly manner, as well as in the time and style which corresponded with the intention of the composer.

Then follows an eye-witness account of Mozart's ability at extemporaneous composition and further attested statements to prove that he had in fact composed those works of music that had been published under his name.

The author then refers to other known child prodigies and cites the case of a John Barratier who was "said to have understood Latin when he was but four years old, Hebrew when six, and three other languages at the age of nine." Reference is also made to a book, *Memoirs of Handel,* by the Rev. Mr. Manwaring, in which it was reported that Handel played the clavicord at seven years of age and composed church services when he was nine. The paper then concludes:

> I think I may say without prejudice to the memory of this great composer, that the scale most clearly preponderates on the side of Mozart in this comparison, as I have already stated that he was a composer when he did not much exceed the age of four.

The statement of the specific objective of the investigation, the description of the various tests and accounts of Mozart's abilities, the discussion of the results, and the conclusions, related to examples cited previously, is exactly the same format used in scientific papers today.

Formal and Informal Communication

The role of journals is sometimes misunderstood because of a failure to distinguish between the formal and informal systems of communication in science. Scientists exchange ideas and information on an informal basis — by private correspondence, during meetings and conferences, at seminars and lectures, and on visits to each other's laboratories. These activities are essential for scientists and for the good of science because they provide a mechanism for the preliminary testing of ideas and for stimulation of new directions of thought. However, science could not progress solely by the use of this informal system. First, the information exchanged is preliminary, still subject to verification, and much of the discussion and interpretation is speculative. Second, the informal exchanges are not recorded, and even if they were, it would be difficult if not impossible to pick out what was definitive. Finally, informal communication is restricted to those scientists who happen to be there; thus, the circulation of ideas is limited.

In contrast to the free and easy character of the informal system of communication, the formal system, publication of papers in journals, has always had some constraints that are designed to maintain order and reliability and to provide for a continuing, accessible record. These constraints can be defined as follows:

- The paper should report a specific, identifiable, advancement in knowledge. To advance knowledge means that the paper must describe something new, and newness implies that it has not been published before.
- The paper should claim no more than can be substantiated; interpretation should not become confused with speculation.

- The paper must be logically consistent, not only within itself, but also within the existing body of knowledge.
- The research reported in a paper must be testable and repeatable by other scientists who are knowledgeable in the subject.
- There must be due reference to previous work upon which the research depends.
- The paper must be available to all scientists. This means that it must be published in a journal that is part of the open literature, obtainable through subscription by any library or scientist.

These constraints, or conventions if you wish, have become an essential part of the scientific process. Their value in preserving order and reliability in the scientific record has been recognized and accepted for more than three centuries. Without these constraints, the scientific literature would be like an uncultivated garden overgrown with weeds—a jungle. An important value of papers, written for a critical audience, is to force the researchers to complete their investigations to the best of their abilities. There is simply no substitute for the carefully prepared, thoroughly vetted paper, published in a reputable journal, as the final step in the research process.

Quality in Scientific Journals

Scientific journals today are still sound and viable in their function as the formal communication system and record of science. At the same time, everyone knows of papers that should not have been published, not necessarily because they were incorrect but simply because they had nothing of any consequence to report. Not only do such papers clutter up the record and obscure the significant work, but they also lead to loss of confidence in the scientific literature and even to cynicism as expressed by such comments as "a library is a record of everyone else's mistakes," and "there are two kinds of scientists, those who write papers and those who read them." Such attitudes may not be widespread but they do exist and represent a risk to the credibility of the scientific record and, thus, to the advancement of science itself. The problem of trivia in the scientific literature arises from attempts to use journals for purposes other than those for which they were intended. Journals are published to communicate and record new science. They are not intended as instruments to amass bibliographies to support advancement, tenure, grant applications, or even Nobel prizes. The pressures to "publish or perish" are well known and present a challenge to the scientific community to increase the significance of qualitative assessments. As far as journals are concerned, the system may be self-correcting by the same mechanism used in dealing with proliferation; i.e., only the significant journals are used. A journal that persists in the publication of a preponderance of trivia can expect to become ignored, lose subscriptions, and eventually suffer a well-earned death. The assessments required to avoid trivia and maintain quality are made by the editors, the key people in the publication of primary scientific journals.

Chapter 2

Editors

The Scientific Editor

The word "editor" means different things to different people. In the context of this book the editor is the person who decides what papers will be published and ensures that those papers are in a form that is scientifically acceptable. Quite simply, this involves making sure that every paper meets the accepted constraints: contribution to knowledge, claims within reason, logical consistency, testability and repeatability, and due reference to prior work. This role as "gate keeper" is certainly the most important function of editors of scientific journals and obviously requires that editors be active scientists. Such editors will not normally have the time, energy, interest, or skills to be involved in copy editing, rewriting, proofreading, printing contracts, subscription fulfillment, or journal finances. These latter activities are regarded here as publishers' functions, for which business managers or managing editors should be responsible, at least for major journals. The distinction between scientific editors and managing editors is essential because neither can properly do the other's job. Most scientists have little knowledge of the publisher's functions, and managing editors, although they may have a scientific background, simply do not have time to keep active in scientific research.

Of course, scientific editors have *some* concerns about the publishing aspects of their journals. They want above all to see their journals appear on schedule and in a text that is both readable and free from errors. They may even be interested in whether their journal is within its budget. However, scientific editors, as defined here, are directly involved in selecting the contents of their journals and only indirectly in the subsequent publishers' functions.

Some other activities that are ancillary to an editor's main role as "gate keeper" include arbitrator, counselor, promoter, and guardian of ethics. Science abounds in controversies, particularly at the frontiers of research. Rather than being a bad thing, the debate and eventual resolution of controversies often contribute materially to the advancement of science. Editors faced with legitimate controversies must decide how to handle them in the best interests of science and the journals. In that capacity editors function as arbitrators. Editors will also have occasion to act as counselors by giving advice, sometimes even solicited, to authors or potential authors about how best to publish their work. The world of scientific publishing is highly competitive, and few if any journals can claim to have a monopoly in any field. Editors may therefore want to promote submissions to their journals in attempts to gain a larger share of important papers. Of course, all editors need to be conscious of the ethics of science and of scientific publication. Should their journals become involved inadvertently in a breach of ethics, editors should know what corrective action to take.

These functions of an editor can be summarized simply by stating that editors are responsible for the scientific standards of their journals. Editors who do not maintain high standards may run the risk of receiving a photograph like Figure 1 with the accompanying letter:

> You are probably concerned with the relative standing of your journal compared to journals published elsewhere. I thought therefore that you would be interested in the enclosed photograph that illustrates a particular strength of your journal. Some other journals were tried for the same purpose but were much less satisfactory.

Editorial Appointments

The appointment of an editor is undoubtedly the most crucial step in the publication of any journal. Editors are involved in definition of editorial scope, appointment of editorial boards, the attraction of good papers, and maintenance of standards. They also represent their journal's main contact point with authors, referees, and readers, and in that context they imbue the journal with a personality. It is no exaggeration to state that a journal's very existence depends upon the quality of its editor: a good editor can make a journal a roaring success; a poor editor can wreck it. Given the importance of editorial appointments, it is surprising how little has been written on the subject. Some interesting questions arise: Who should be an editor? How are editors selected and what guidelines, if any, are used? Why would anyone want to be an editor? How long should an editor's term be? This chapter attempts to answer these questions in a general way. It is dangerous to be too specific or dogmatic on the subject of editors because editorial work involves dealing with people

Figure 1

and is therefore personality dependent. For that reason there will always be some exceptions to any guidelines or recommendations.

Anyone involved in the selection of editors should give very serious consideration to the qualities desired. Observation of good editors, past and present, shows that the following guidelines are applicable. Good editors should

- have an established record of published research
- be currently active in research
- be reasonably well organized
- have tact, diplomacy, and good judgment
- have a sense of humor
- be at an appropriate stage in their careers

An Established Record of Research

This point relates to the credibility and visibility of the editor. Authors and referees feel more comfortable with an editor whose name they recognize and whom they respect for a significant record of research. This does not mean that prospective candidates should be measured against each other by the size of their bibliographies or number of citations. However, there should be evidence that the people being considered are solidly established in their field and have knowledge and ability. The phrase "this scientist has an international reputation" is an overworked cliché, but in its proper sense it represents a desirable qualification for an editor. Some indicators other than publications are also pertinent. Does the person participate in symposia, conferences, and seminars at home and abroad? Have there been reciprocal visits with scientists in other laboratories? Is the person consulted frequently by others in the same field? Has the person won any awards, national or international, including election to academies or honorary status in societies? The answers to *all* of those questions need not be positive (the granting of awards can be capricious), but there should be some indication that a prospective editor is known in the peer group that is advancing that particular field of science.

Currently Active in Research

In any search for a new editor someone will almost certainly come up with a suggestion such as, "Jim Longtooth worked in this field for 40 years and retired 2 years ago. He has plenty of time, is in reasonable health, and would do a good job." This sort of suggestion should be regarded with extreme caution because, with all due respect to Jim Longtooth and his 40 years of undoubtedly good research, such a person is most likely to be out of touch after 2 years of retirement. There is, of course, the odd exception. A very few retired scientists continue to read the literature, attend meetings and conferences, and may write reviews. Such activities may serve to keep a scientist reasonably well informed, but the current awareness of editorial candidates in this category should be examined very carefully. Scientific advances come quickly these days, and anyone who stops research activity becomes outdated at the same rate. It is important for an editor to know where new, exciting work is being done and by whom, and to retain an appreciation of quality in experimental work. Furthermore, authors want to be judged by people who are playing the game. All of this indicates that the best candidates for editors will be found among those scientists who are currently active in research.

Well Organized

Everyone knows some (perhaps many) scientists whose laboratories, offices, and even their very lives, look as though they are caught up in a hurricane. They often are excellent scientists and, indeed, their seemingly chaotic life style may derive from such intense concentration on scientific problems that all else is unintentionally ignored or forgotten. Such people ought not to be editors, and it is important to note that this is in no way a criticism of their scientific or intellectual capabilities. A journal in the hands of such a scientist runs the risk of having manuscripts, reports, and correspondence mislaid under reprints or lab notes, or piled up and neglected while the editor pursues promising and important experiments. In addition, such a scientist, though perhaps professing a willingness to take on editorial functions, will quickly come to resent the job as an intrusion into an already cluttered scene and a further limitation on the time available for research.

This does not mean that prospective editors should be known to keep spotless desks, or labs so clean that it is obvious no work is going on there. There is a saying, "an untidy desk is the sign of a scientist, a clean desk the sign of an administrator," and editors should certainly be scientists. However, the operation of an editorial office, organization of editorial boards, and maintenance of the flow of manuscripts requires a modicum of managerial skills. Prospective editorial candidates might therefore be expected, at the least, to know what time dinner is served, to be able to catch a flight on time (without losing the tickets), and to remember that mail needs to be read and answered. A good editorial assistant can keep the office running smoothly and deal with routine matters, but items that require the editor's attention must still be handled on some sort of organized schedule to avoid delays and backlogs. Clues to the organizational abilities of prospective editorial candidates can be obtained by examining their performances in extracurricular activities such as service on national or institutional committees, active participation in the affairs of a society, or organization of meetings. The old adage that "if you want something done, you ask a busy person" holds as true for editing as for other activities. Busy people are usually those who can successfully organize their time.

Tact, Diplomacy, and Good Judgment

The need for these qualities in a journal editor should require little explanation. With the availability of so many alternative routes to publication, no journal can afford to court the wrath of authors or referees by having an insensitive, rude, autocratic editor who has an abrasive personality. Editors certainly must be decisive, but they should be able to explain their decisions in language that is at least polite. A rejection that reads, "From considering these reports it is clear that there is nothing here worth publishing; I wish never to see this paper again," may be contrasted with, "In view of these reports I regret that we cannot consider this paper further for publication. Perhaps you will find the referees' suggestions to be useful in your future work." The message is the same, but to which journal will the next paper be submitted? Of course, subsequent papers from the same authors may well be very good, particularly if the authors are young and just learning the ropes, with benefit to the second editor's journal. Some few successful editors are highly autocratic and, by force of personality, scientific stature, and the prestige of their journals, can get away with things that would spell disaster for anyone else. The acknowledged success of those few autocrats does not justify imitation in situations where there is so much variability in personalities and circumstances. The safe procedure is to seek editorial candidates who are firm and decisive but also polite and considerate in dealing with people.

Tact is simply a matter of consideration for the feelings of others: authors, referees, associate editors, editorial assistants, copy editors, and anyone involved with the journal. Once an editor is immersed in the daily flow of manuscripts across the editorial desk, it is easy to forget that there is a real, live, human being behind each of those pieces of paper. Authors feel parental towards their papers and need to know that the editor is keeping track and paying attention. Referees are doing unpaid favors for the journal and need to know that their efforts are appreciated. Everyone responds better to a carrot than to a stick, and a "thank you" for a job well done can reap hidden benefits. Editors who are sensitive to the feelings of others will receive help and cooperation that will make their editorial work both pleasant and efficient.

Diplomacy is defined as the art of negotiation, usually involving the recognition of possible compromises and ways to achieve them. A diplomatic editor does not view everything as black or white but recognizes that there are various shades of gray in between. This does not mean that editors should frequently get involved in prolonged negotiations between authors and referees. However, the editor should be able to see and negotiate a compromise when an author has accepted the substance of a referee's criticisms without necessarily having followed the recommendations to the letter.

Good judgment is largely the application of common sense and does not necessarily correlate with intelligence or scientific skills. It is poor judgment to make hasty decisions based upon incomplete information, and it is also poor judgment to delay or waffle on a decision once all the information has been considered. Clearly, editors must be decisive, but they must also be sure that their decisions are sound and made only after careful consideration of the best available information. Several decision points occur in the editorial process:

- selection of referees (Were those chosen the best that could be found or would a more careful search reveal better ones?)
- receipt of referees' reports (Are the reports reasonable and do they show adequate knowledge of the subject? Do the reports require editing? Should a third or fourth referee be consulted? What directions should the editor give to the authors?)

- receipt of authors' revisions (Are these acceptable? Should further advice be sought from the original referees, or from a new referee?)
- final decision to reject or accept

At each of those points an editor must decide what to do. Thus, in a search for prospective editors, it is valuable to know that a candidate has demonstrated the ability to make sound judgments and stick to them.

A Sense of Humor

There are several reasons why this personality trait should not be regarded as a trivial requirement for an editor. Perhaps the most important is that a light touch can often avoid, or at least soften, potential disagreements among authors, referees, and editors. An unwelcome message will often be more acceptable if delivered with a bit of humor. For example, there was an editor who had to relay a particularly vituperative referee's report to an author who was himself rather testy at times. The editor's letter read, in part, as follows:

> You will probably foam at the mouth when you read referee II's report. It is written in such a way that I could not edit out the abusiveness without also running the risk of losing pertinent criticisms. May I say that we have used this referee before and his reports are not always like this. Perhaps we caught him on a bad day (hangover, accident, fire?). In any case, could I ask you to read around the abusive language and focus on the valid scientific points that may be helpful in preparing an acceptable revision.

Some weeks later the editor received the following reply from the author:

> In my opinion referee II, if not an idiot, is really rather naughty. "Methinks he doth protest too much. . . ." I do not know whether he is one of the authors of the dozen or so papers that have been written on my original measurements but he has misunderstood the point of our paper. Perhaps this is due to our schoolboy English or even, dare I suggest it, to his schoolboy level of comprehension? However, though I definitely disagree with his statements, I have decided to do some more measurements before we resubmit the paper in altered form.
>
> Yours sincerely,

Please note the final sentence of that letter; editor's mission accomplished.

Of course, there are other techniques for dealing with abusive language. For example, the editor can edit, or rewrite, the report and have it retyped before transmission to the authors. However, at least one editor has been known to use secretarial correction tape to cover up nonessential vituperation even if grammatical continuity is lost. The author is then told in the covering letter that the referee's report was "Watergate taped" to limit the comments to the scientific essentials. The result is usually that the author is more curious than angry and, in attempts to fill the gaps, is sometimes carried further than the original comments intended.

Another editor tried to temper a rejection with a little humor as follows:

> This journal wishes that all of the papers it receives were like sirloin steaks; we know that this is a world mainly of Big Macs, but we simply are not in the business of competing with White Castle. The current manuscript seems just a little bit extra on previous work and, in order to qualify as a meal, would have to be bought by the bagful. Please send it somewhere else.

After this letter was sent, the editor had second thoughts that it was perhaps more tactless than funny and wrote a letter of apology for his lack of sensitivity. However, the apology crossed in the mail with a letter from the author that read as follows:

> You bastard, you are correct but you don't have to say it. The paper has been thrown in the round basket.

Again, editorial mission accomplished and, apparently, no hard feelings.

Finally, a sense of humor provides a great survival mechanism to help editors withstand the slings and arrows of outraged or outrageous authors. Editorial work is not for the thin-skinned and in particular not for those who take themselves too seriously. People who can laugh at themselves realize that nobody is infallible and are also tolerant of mistakes by others; these are essential qualities for survival as an editor.

Appropriate Stage in Career

Potential editors should be enthusiastic and committed to a journal's interests. Enthusiasm is required because without it the editorial functions will not receive the attention they deserve; commitment is necessary because too frequent changes in editors are disruptive to the continuity of the journal operation. The key to finding that enthusiasm and commitment is to assess any competing motivations that might be present, and take precedence, in potential candidates. For example, there are scientists who, at the most productive stages of their careers, resent any activity that detracts from their research. Such people will shun committee work, administration, and society executive positions, and they will be certain to give editorial work a low priority. Similarly, the scientist who is in hot competition for tenure or advancement to a more senior position cannot be expected to give precedence to journal business. None of this means that potential candidates for editorial positions should be ossified or immovable. Indeed, the qualities that make a good editor may also make that person an attractive candidate for a variety of senior positions in science. The prime distinction is motivation; the better editorial candidates will be found among those who are sought after, not those who are seeking. In sum, a search for an editor will probably yield better results if focused on scientists who are productive and happy in a secure position, not involved in a competitive situation with respect to their career advancement.

The Selection Process

The Selection Committee

Each journal will presumably have some body associated with it — a publication committee, an editorial committee, or a management committee — to which the publisher delegates responsibility for selecting an editor. The members of this committee should be made cognizant of the guidelines for selection of editors: that selection requires careful and thoughtful assessment and is *not* a popularity contest.

Confidentiality

The committee members should also be advised to keep their discussions and any individual inquiries on a confidential basis. In this day of open democracy and freedom of information, a recommendation for confidentiality may seem anachronistic, but there are good reasons for it. First of all it is clear from the guidelines that some superb scientists are not at all suited to be editors. Should it become known that such people were rejected as possible editors, that decision will most certainly be misinterpreted because the scientific community is so accustomed to evaluations based solely on the quality of the science. By failing to observe confidentiality, selection committess run the risk of needlessly and mistakenly giving offense to some excellent scientists who may be strong supporters of the journal as both authors and referees. Similarly, the selection committees may come up with a list of candidates any of whom would be excellent editors. The names must nevertheless be ranked, if for no other reason than to provide an order of approach, and such a ranking literally invites misinterpretation. Thus, confidentiality is recommended in the selection process, not for any sinister reasons but to prevent unjustified offense being given or taken as a result of misinterpretations.

Mechanism

The selection committee, using the guidelines, should generate a short list of say four or five names of potential candidates and rank them in order of preference. Suggestions and information may be obtained from the committee members themselves, from the retiring editor, from past editors, and from members of the editorial board. At that stage candidates should not be approached to see if they would have time to be an editor or if they would accept. Any such inquiry would not only breach the confidentiality of the process but also run the risk of generating a premature negative response. Good potential candidates are busy people who will come up with manifold reasons why they could not possibly become editors, particularly if they are given time to think about it without being given the full picture. It is far better for the committee to construct its short list, rank the names according to the guidelines, and then have the chairman make the first approach, preferably by telephone. In this way the chairman can be prepared to outline the positive aspects of editing, of which the candidate was probably not aware, to explain the selection process, and to emphasize the very real contribution that a scientist makes by serving a term as editor. The chairman should also be prepared to explain what assistance the journal provides to its editors and to answer any questions that the candidate might have. By this procedure, candidates can make a decision based upon full information, and it will be a better decision than one based upon a preliminary inquiry, "Do you think you might have time to edit the journal?"

The Acceptance Process

Candidate editors who have thought about the positive aspects of editing, and are attracted by the idea, should seek further information from their home institutions, the publisher, and the retiring editor before making a final decision.

Home Institution

Prospective editors should find out whether their home institution would welcome or disapprove the appointment of one of their staff as an editor. Most institutions consider it an honor to host an editor; they see the very real benefits in the closer contact with the scientific community, and in increased visibility from the editor's address on the masthead of the journal. In return for these benefits the host institution may be asked to provide space for an editorial office and possibly to establish an account within its administrative structure for the financial operations of that office. Perhaps the host institution would prefer to enter into a contract with the publisher and establish the relationship in a formal way. These are all items for investigation and negotiation by prospective editors and the publishers of their journals, but they should be settled before final acceptance of an editorial appointment.

Publisher

Prospective editors should have an understanding of what may reasonably be expected of the publisher and should seek information to clarify any uncertainties. Arrangements between publishers and editors vary considerably and may be negotiable. Reasonable guidelines for prospective editors to follow in such negotiations are

- the publisher should cover the operating expenses of the editorial office either through a contract with the host institution or through a separate account with the editor
- the publisher should provide for secretarial assistance to deal with all routine matters that do not directly involve the review and assessment of manuscripts

Thus, the publisher might be expected to cover such operating expenses as office supplies, stationery, postage, and telephone, and to provide equipment such as typewriters, word processors, photocopiers, recorders, and filing cabinets. These items may be either rented or obtained by direct purchase for transfer to succeeding editors. To have enough time to maintain their qualifications as active researchers, editors need assistance with the routine matters of the editorial office: someone to answer the telephone, maintain the records, send and receive mail, type the letters, keep the files in order, and manage the follow-up system. These duties properly belong to a secretarial assistant whose salary should be paid by the publisher.

Although the above guidelines are valid for all journals, the magnitude of the assistance to the editor from the publisher, and the circumstances of its delivery, may vary considerably from one journal to another. Some small journals are edited as a labor of love; the editorial office may be the editor's study at home, and most of the work is done on weekends or in the evenings, with perhaps some assistance from another member of the family with typing and filing. Publishers with such devoted editors are very fortunate and might well show appreciation by a cash honorarium, by paying for the trip for the editor to attend a scientific or publishing meeting, and of course, by covering the operating costs of the editorial office.

With larger journals the form of assistance from the publisher depends upon the editorial system used. A fully centralized system, in which the editor handles all correspondence, obviously requires a full-time editorial assistant. In decentralized systems the senior editor in charge of the main office requires a full-time assistant, but, depending on the work load, the subeditors or associate editors may not. Various options are available. The editors and publisher can agree that a full-time editorial assistant will only be provided to an individual who deals with more than a certain number of manuscripts per annum (100 to 200?). For smaller work loads, a part-time assistant might be considered, paid either an hourly rate or a fee per manuscript handled. For still smaller work loads, the subeditors or associate editors involved may be able to utilize their regular secretarial services with payment again by the hour (perhaps overtime) or per manuscript handled.

Although all of these items of assistance from the publisher are negotiable, the main considerations are that the publisher should cover the costs of operating the editorial office and provide sufficient and reasonable administrative help so that editors can maintain their research activity.

Retiring Editor

The final source of information for candidates who are considering acceptance of an editorial appointment is, of course, the retiring editor of the journal, the only person who can truly define such things as the work load, the time involved, the rewards and satisfactions, and the qualities of associate editors. Journal statistics are of interest as well as a description of the editorial system and any specific problems that may have arisen with the publisher, associate editors, managing editor, copy processors, or printers. All of this information serves to let the prospective editor know just what the job entails and may also indicate areas where improvements can be introduced. Even if the prospective editor has already decided to accept appointment before consulting the retiring editor, an information briefing is essential in arranging the smooth transfer of editorial duties.

How Long a Term?

The editorial life-span of individuals is variable, but the following pattern can perhaps be regarded as typical. During the first year, a new editor is busy learning the job, responding to the challenge, and trying hard not to make mistakes. With the confidence of a year's experience, the editor then develops some new ideas to improve the journal and implements those that are feasible during the second and third years. In the fourth year the now well-experienced editor will consolidate the operation, streamline procedures, and possibly make some adjustments in the editorial board. With the excitement of the first two years gone and with new ideas implemented in the third and fourth years, the fifth year can become very much a holding operation. At this point some editors, who are and should be productive scientists, may actually come to resent the time required for those editorial functions that they approached so enthusiastically in the first three years. When that happens, it is time for a change. If this timetable bears any relation to reality, then a reasonable term for an editor is four to five years. An editor will not make much of a contribution to a journal during a term of less than four years, and changes in editors more frequently than four or five years can disrupt the continuity of the journal operation. An editorial commitment of four to five years is therefore in the best interests of the journal and is not so long as to discourage good candidates from serving a term as editor.

Of course, the terms should be renewable to permit journals to retain those editors who fall outside the normal pattern, who retain their enthusiasm for editorial work and continue to do a good job for 10 or 15 years. Good editors are assets to be treasured and, when they are willing to continue, should not be dismissed because of some arbitrary deadline. However, the term dates provides review points for both editors and publishers. The editors can decide whether or not they wish to continue; publishers can assess whether fresh blood is indicated. Publishers can have a number of reasons for wanting to change editors after one or two terms. Perhaps a journal has fallen into a comfortable rut, still good but rather pedestrian. Under those circumstances an appropriate new editor can sometimes give the journal a new image, alter the editorial scope, and bring about a revitalization. Editors who start neglecting their duties, for whatever reason, should be replaced, and quickly, before irreparable damage is done to the journal. Nothing is more harmful to a journal's reputation with authors than to have long, unexplained delays in refereeing, lack of communication with the editor, and repeated evidence of inappropriate selection of referees. Editors are, and should be, accountable to authors, to the publishers of their journals, and to their colleagues on editorial boards, all of whom have the best interests of the journal at heart. There will always be a few authors who insist that they have been treated unfairly and may even voice their complaints to the publishers. Of course, every complaint about anyone connected with a journal must be taken seriously. Authors are allowed to express their strong feelings of mistreatment. Editors are not allowed either to ignore these feelings or to express their personal frustrations; unlike authors, they are expected to be perfect. Publishers should therefore support and reinforce their editors in the absence of any concerted evidence of dereliction of duties.

The Joys of Editing

Why would anyone want to edit a scientific journal? The common conception that editors are unloved, overworked, and underpaid has a certain validity. Disgruntled authors abound, and many would subscribe to the opinion that "editors are a low form of life — inferior to the viruses and only slightly above academic deans."[5] Somebody once said, "If you ever see an editor who pleases everybody, he will neither be sitting nor standing — there will be a lot of flowers around him."[5] Certainly editors have to work hard, just to keep pace with the daily flow of manuscripts, and a lot is done at night after normal working hours (although not in the dark as some authors might claim). Few editors are paid, and those who do receive small honoraria may be expected to use them to meet certain costs of the editorial office. Why then would any scientist take on a job that seems to be thankless, that is bound to make them unpopular with at least a number of their colleagues, that is filled with frustration, that will add hours to their working day, and that provides no obvious material benefit? Despite these apparent drawbacks, there are many scientists involved in editorial work as editors and associate editors. Moreover, good candidates continue to be found when those positions rotate. What motivates these people: lust for power or prestige — they cannot all be masochists? Most editors, when asked why they edit a journal, will give a simplistic answer such as "because I enjoy it" or "because it is fun." However, the fun and enjoyment arise from a number of very positive factors.

Variety and Enrichment

One reason why busy people always seem to have time to squeeze in one more job is because they enjoy the variety of different activities. Editorial work is quite unlike the other activities in which scientists are involved such as research, teaching, consulting, or even administration. In agreement with the old adage that a change is as good as a rest, the time devoted to journal editing is not drudgery but a very pleasant change of pace. There is also variety within the editorial process itself. An editor never knows what the next mail will bring by way of new submissions, revisions, reports, and correspondence from authors and referees. The paper flow may sometimes be intimidating or challenging; it is seldom dull.

Research scientists who are also editors find that the two activities enrich and support each other. A laboratory scientist brings to the editorial desk an appreciation of quality and standards in experimentation and interpretation that is invaluable in the review process. Editors find that the critical faculties and judgments, required by editorial work, carry over into their own research which is thereby improved. Editors, in their research, naturally and perhaps subconsciously, become reluctant to do work that is below the general standards of their own journals. Also, editors acquire an up-to-date knowledge of their fields from reading submitted papers, the papers of potential referees, and papers that may be cited in referees' reports. It is, of course, unethical for editors to make direct or specific use of unpublished work of which they became aware through their privileged positions. However, in the general sense, editors simply become more current than most scientists, with consequent benefits to their research.

Knowledge of Scientific Community

Editors acquire a knowledge of the international community in their fields of science that simply cannot be obtained in any other way. Not only are editors aware of the latest developments, but their innumerable contacts with so many authors, referees, and associate editors allow them to develop a broad and unique perspective of an entire field, including the personalities involved. An editor's correspondence with authors and referees is often informal and frequently revealing of personal characteristics. Although that correspondence is confidential, never to be cited, quoted, or discussed, astute editors will soon form, quite unsystematically, a reasonably accurate picture of people whom they have not even met. Thus, editors are veritable fountains of knowledge about who is doing what and how well in their fields. Such knowledge is of benefit in many useful and legitimate ways. For example, editors can give reliable and knowledgeable suggestions of people who could be consultants, directors, collaborators, or new staff members. Editors also know the strengths and weaknesses of various laboratories and can respond to such questions as: "Where would be a good laboratory to spend a sabbatical year to pick up a particular technique?" or "Can you suggest a laboratory that might have the expertise and interest to collaborate on a certain project?" Of course, much of this knowledge is exchanged in the invisible colleges of each field of science in the normal course of events, but editors are certainly among the best-informed members of those groups.

Satisfaction

Most editors probably get their greatest rewards simply from the satisfaction of doing a good job. That satisfaction arises in many ways. It is satisfying for editors to see their journals grow in size and stature, and to know that the papers published represent solid contributions to science. It is a source of pride for editors to know that their journals are highly regarded, are cited by others, and form a significant part of the literature of science. Editors also derive satisfaction from seeing their new ideas introduced, be it a major change in the editorial scope or direction of the journal or a minor modification in format to make the journal more attractive. All achievers enjoy seeing the results of their efforts, and editors are no exception.

However, the greatest satisfactions arising from editorial work are to be found in the human relations with authors and referees. New editors will quickly find that the vast majority of scientists are not only reasonable people but that they have high standards, are exceedingly generous of their time, and very pleasant to boot. Referees really do try to be helpful and will go to great lengths to make a paper suitable for publication as long as the science is sound. An example of that attitude was summed up in a letter from an arbitrator of a paper that had been through a couple of revisions without satisfying the referees. The arbitrator wrote:

> If I may offer the author free advice (which is probably worth approximately what you pay for it), I would say don't feel injured by erroneous referee's reports. Referees are better than average readers. If you can't reach them, your paper is in trouble even if your science is sound.

Another example from a referee who obviously found a paper acceptable but wanted to be helpful:

> I found this paper to be informative, well-written, and to have the theory firmly connected to experimental observation, a feature in sad decline in much of the current literature. The four comments listed are intended only as suggestions to help with the appreciation of the contents by slow learners such as the present reviewer.

Most authors recognize that referees try to be helpful and, more frequently than might be expected, editors receive letters like the following:

Dear Editor:

Thank you very much for the referees' comments which are both very perceptive and very helpful.

Dear Editor:

Although the referees' reports were in some respects irritating, in many cases they make justifiable comments. We have responded to most of these and without a doubt the text is now more coherent, and less verbose. We are grateful for the efforts of the referees.

Dear Editor:

Did you ever hear the final story on a controversial paper I submitted about a year and a half ago? After discussions about the referees' reports, your Associate Editor agreed to seek the opinion of a third referee as arbitrator. This referee found a major error in the theoretical part, which had escaped 4 authors, 2 previous referees and the Associate Editor! My gratitude is boundless; I was saved from a very embarrassing situation.

Letters such as these, and there are many, bring joy to the hearts of editors and make them realize that they represent more than just a relay station.

Finally, editors can gain enormous satisfaction from guiding good young scientists in preparing their work for publication. Perceptive and sympathetic editors can help young scientists to redraft their papers, or encourage them to pursue a promising line of research. Sometimes some guidance about publishing strategy is beneficial: when to publish a Communication, how to avoid undue fragmentation, and how to divide a program of research into substantial, publishable papers and still retain continuity. In these ways, editors can have a real and lasting influence on their fields.

Humor

Scientific publication is normally a serious business, as it should be. Happily, however, there are some scientists who, while very serious about science, do not take themselves seriously at all. Dealings with such people can brighten an editor's day and may even lead to some harmless pranks. One author, having received very mild criticisms of a paper, returned a revised manuscript with the following comment:

The referees of our manuscript were indeed gentle — so much so that perhaps you should hesitate about sending them any further manuscripts on the grounds that they do not meet the standards of toughness, even bloody-mindedness, normally required by this journal.

Another manuscript received only minor criticism but a spelling error — "taught" was used instead of "taut" — was caught by a referee. The author, in returning a revision, wrote:

> Dear Editor:
>
>   It's nice to get one through unscathed, more-or-less, now and again. If you ever see the referee concerned, thank him from me for pointing out the tort of confusing taut with taught; it was very norty of me and I am mawtified!

Some authors, rather than becoming abusive, find more elegant (and more effective) ways to make a point, as shown by the following letter from an author whose manuscript was long delayed in the review process.

> Dear Editor:
>
>   In regards to dilatory reviewers, I would like to quote from Anthony Trollope's Autobiography — "The punctual who keep none waiting for them, are doomed to wait perpetually for the unpunctual. But these earthly sufferers know that they are making their way heavenwards — and their oppressors their way elsewhere. If the former reflection does not suffice for consolation, the deficiency is made up by the second."

Authors sometimes play tricks on editors, who are easy prey because they know that authors normally regard publishing as a deadly serious activity. One group of scientists, taking the English folk song "Widdecombe Fair" as their inspiration, added the last words of the chorus, "uncle Tom Cobbley and all," to the list of authors on one of their papers. The prank escaped the notice of the editor, two referees, two copy processors, and the printer, and the paper was published[20] with the authors' names as submitted, much to the chagrin of the editor. In a similar vein, another author was informed of an editor's rule that the word "we" should not be used in a paper with a single author. Rather than change the paper to the impersonal, the author simply added the name of his cat as an author.[26] There is also the true story of a senior scientist who persuaded one of his younger colleagues, who was quite adept with words, to write the abstract of their next paper in blank verse as a joke on the editor. The paper, with the abstract in blank verse, was accepted and published,[15] whereupon the senior scientist shot off a letter to the editor reproaching him for not spotting this joke. However, the editor responded that he had indeed noticed the blank verse but let it stand because he could not see how the abstract could be improved. The verse form of that abstract is reproduced here with the kind permission of the authors.

FARADAY FANTASY

The exchange between gaseous chlorine and crystalline sodium
Chloride was used to enable diffusion parameters
Of the negative ions to be measured with equal precision
For both the intrinsic and the impurity ranges.
The results showed a difference of zero point six eight electron
Volts in the slopes of Arrhenius plots of diffusion
Coefficients between the two ranges, which must be compared with
Half of the energy needed for thermal creation
Of a vacancy pair, for which values from one point nought six to
Zero point nine three eV have been quoted from theory.
The discrepancy can be accounted for on the assumption
That the impurity centres and anion vacancies
Interact at low temperatures forming immobilised complexes
Which require zero point six six eV to dissociate,
While the energy of activation of anion migration
Is unity decimal two eight electron volts.

A few occasions arise when editors may wish to collaborate, or at least cooperate, with authors in publishing a bit of clever humor. The following papers might be cited as examples: "Congressane,"[4] a report of the synthesis of a polycyclic hydrocarbon, the structure of which was used as a logo for a meeting of the International Union of Pure and Applied Chemistry; "The Stereochemistry of Octahydrohexairon: A Molecular Raft,"[25] a humorous spoof on conformation analysis of organic compounds; "Psychochemical Symbolism,"[18] a description of the phallic symbolism of organic chemical structures; and "Studies on a New Peerless Contraceptive Agent: A Preliminary Final Report"[13] — in this amusing shot at the drug industry, the new contraceptive was named "armpitin." While nobody would suggest that journals carry a regular comic page, everybody must surely appreciate a harmless joke that is carried off with wit and style.

Offbeat Submissions

Whether or not offbeat submissions should be included among the joys of editing is difficult to judge because they arouse different emotions among editors: anger, irritation, pity, pathos, or amusement. However, they do introduce variety and a diversion from the mainstream of papers.

Offbeat submissions come from pseudoscientists, would-be scientists, cranks, crackpots, or retired scientists who are no longer fully up-to-date. All such submissions have characteristics that make them easily recognized:

- The paper will come from a private rather than an institutional address.
- The subject matter will be broad, usually covering a whole field, and will contradict the well-established consensus of knowledge in that field.
- There will be little, if any, experimental work and what there is will be totally lacking in controls or evidence of reproducibility.

Just how these submissions are handled is a matter of editorial choice. Undoubtedly, some editors consign them directly to the round basket, others may simply return the manuscript either with no reply or a simple statement that, "this paper is not suitable for publication in this journal." Still other editors, if the paper is at all intelligible (many are not), may pass it to a colleague for a very brief review that emphasizes the contradictions of the consensus of knowledge in the field. Even those editors who adopt the last procedure, as being more polite and humane, should be clear and firm in their rejections. To do otherwise is to invite an endless stream of correspondence and further meaningless submissions; authors of this bent are nothing if not persistent.

Chapter 3

Editorial Boards

The early scientific journals functioned very well with just one editor, to whom manuscripts were submitted, who selected referees and accepted or rejected the papers. With the growth and specialization of science, it became more difficult for any one person to be knowledgeable in all topics covered by a journal so editors designated specific assistants, the first editorial boards. Nowadays, there is hardly a journal published that does not have a list of names on the masthead under a title such as Associate Editors, Editorial Board, or Editorial Advisors. These groups of editorial aides are used in different ways by different journals depending partly upon the subject matter of the journals and partly upon editors' styles.

One type of board can be dismissed quickly: the non-working editorial board that may be likened to a vanity press. Eminent scientists are invited to become members of an editorial board, with a promise that little work is involved, and will usually accept because such invitations are regarded as recognition of stature in a field. Scientists who accept those appointments will have their names listed on the mastheads but, in some instances, may never hear again from the journals or their editors. In fact, contacts may be so infrequent that journals are not aware of deaths and continue to carry on their mastheads the names of deceased members of the board. Obviously, journals that have non-working editorial boards must depend solely upon their editors to conduct the review process and to determine policy. The masthead list of prestigious names is intended only to enhance the stature of the journal in the eyes of potential subscribers.

The following description of editorial systems concerns editorial boards that actually do something, even if it be no more than give advice to the editors. The intent is not to recommend one system as being better than another, but rather to outline the options available and to indicate their advantages and disadvantages. There are some general principles that apply to the formation of all editorial boards: these are matters of size, selection, and term of service. The differences among editorial systems arise from the extent to which the review process is centralized or decentralized.

General Principles

Size

To fulfill its proper functions of advice and assistance, an editorial board should supplement and reinforce the editors' expertise. The actual number of members will reflect the breadth of the subject matter covered by the journal because editors will want assistance in all specialties that are represented by a significant number of papers. Thus, general journals, such as the *Journal of the American Chemical Society*, the *Journal of Biological Chemistry*, and the *Journal of Physiology*, require larger editorial boards than more specialized journals such as *Steroids*, the *Journal of Cyclic Nucleotide Research*, and *Neurotoxicology*, simply because the general journals cover a wider variety of topics.

The need to spread the work load is another factor in determining the size of editorial boards. Clearly, an editor of a large journal, with say 500 or more submissions per annum, will require more assistance with the review process than an editor of a small journal with only some 50 to 60 papers submitted each year.

The size of editorial boards varies considerably from one journal to another, with most of them falling in the range of 10 to 50. For working boards, possibly the most useful guideline for editors is, "small is beautiful." Editors are well advised to strive for the minimum number of associate editors consistent with an equitable distribution of the work load and representation of all specialty areas covered to any significant extent by their journal. To go much beyond the minimum requirements invites inefficiency, loss of contact, and even conflicts. Small editorial boards are easier to manage and tend to develop stronger interests in their journals.

Selection

Editors should be free to select their own editorial boards and should not have people imposed on them by scientific societies, executive committees, or the like. After all, the editors are the ones who must manage and work with their boards, and it is important that they be able to assemble a compatible group. When editors resign, the new editors will inherit editorial boards that were established by their predecessors. The new editors may well be satisfied with the old editorial boards, but it should be clearly understood that changes can be made if desired.

The main concern of editors in forming editorial boards will probably be to obtain the supplementary expertise needed to help with the review process. The first step is to compile a list of the topic areas covered by the journal and to put three or four names of possible associate editors under each topic. Those names may be drawn from lists of referees and authors, or simply from the editor's memory. Generally, an editor will want as associate editors those scientists who have shown an interest in the journal as either authors or referees. However, appointments to editorial boards can also be used to indicate a change in direction or to attract the interest of scientists who have not previously been associated with the journal.

Once the lists of names, by topics, have been prepared, editors will need to know or find out something about the individuals before making a final selection. In many instances, editors will have personal knowledge of the various candidates based upon their performance as authors or referees; if not, a few discreet inquiries would be in order. The qualities sought in associate editors are much the same as those for an editor although the demands are not as great. Associate editors should have shown evidence, by their performance as authors and referees, that they have an understanding of how the review process is supposed to work. Extremes should be avoided. Editors do not need assistants who, perhaps in a burst of enthusiasm, will attempt to rewrite every paper sent to them. Nor do editors need people who make snap judgments and who will not take the time to give carefully considered comments. In short, good judgment should be near the top of the list of qualities for members of an editorial board. Other factors are stature and knowledge in the field, an appreciation of deadlines in the review process and an ability to meet them, and a compatible personality. Good communication and agreement on procedures are essential to the smooth management of editorial boards. Inflexible individuals who cannot adapt, but insist on doing things their own way, can quickly wreck the system. Of course, editorial boards also serve to train and test potential editors. A common progression is from referee to associate editor to editor, based upon performance at each level. Thus, someone who performs well as a referee for a given journal may expect to become an associate editor at some stage. The best associate editors are obvious first-choice candidates to become editor because of their proven abilities and knowledge about the journal operation.

Terms

Editors will find that a defined term for associate editors is a useful management tool. Potential candidates are more likely to accept appointment to an editorial board if they know it is for a set period. Appointments can be staggered to provide for continuity and made renewable so that good people can be retained. A term of two or three years gives a reasonable opportunity for turnover and the injection of fresh enthusiasm where required.

Editorial Systems

Table 1 shows the main editorial systems listed in order of decreasing centralization. There are some minor variations, and the more significant ones will be referred to in the following discussion of each system.

Different Systems

System A is the most fully centralized system. The editor receives submitted papers, selects referees, dispatches the manuscripts, receives and assesses referees' reports, and does all correspondence with authors. In this system associate editors or the editorial board serve as problem solvers. The editor may want to use an associate editor as a third referee when necessary or as the source of a quick, reliable report when one of the original referees has failed to respond. Associate editors may also be called upon as arbitrators on difficult papers and as advisors on matters of policy and editorial scope.

System B is a slightly decentralized version of system A in that the editor delegates, to associate editors, the selection of referees for papers that are specifically within their areas of competence. Usually, the associate editors will dispatch the manuscripts to the referees but the reports are returned directly to the editor for assessment and subsequent action. A minor variation of this system is one in which associate editors simply inform the editor of the referees selected and manuscripts are then dispatched from the editor's office.

System C involves associate editors to a much greater extent than either system A or B. In system C, the editor receives submitted manuscripts but then distributes them to associate editors according to subject matter. The associate editors select the referees, dispatch the manuscripts, receive and assess the referees' reports, and send all of this back to the editor together with recommendations as to the disposition of the papers. The editor therefore has two referees' reports and an associate editor's assessment, based on those reports, as assistance in making decisions. It should be emphasized that under this system editors still have the final word and do more than simply relay the referees' reports and associate editors' assessments. The editors must read the reports, the associate

Table 1   Editorial systems

System	Manuscripts submitted to	Referees selected by	Reports assessed by	Correspondence with authors
A	Editor (1)	Editor	Editor	Editor
B	Editor (1)	Editor and Associate Editors	Editor	Editor
C	Editor (1 or 2)	Associate Editors	Associate Editors	Editor
D	Coeditors (4–6)	Coeditors	Coeditors	Coeditors
E	Associate Editors (30–35)	Associate Editors	Associate Editors	Associate Editors
F	Central secretariat	Editors or Associate Editors	Editors or Associate Editors	Central secretariat

editors' comments, and the paper in order to formulate specific directions to the authors.

In system D, the subject matter of a journal is divided into four to six major sections with a coeditor in charge of each. The coeditors then operate their individual sections almost as minijournals and are responsible for all aspects of the review process for papers submitted to them. If warranted, each coeditor could be assisted by a small group of associate editors. This form of decentralization is in fact a combination of four to six centralized subsystems and requires some coordination and designation of authority. There should be someone who collects the accepted papers to be passed for printing, who knows the number of papers in the works with each coeditor, who is authorized to settle disagreements, and who is ultimately responsible for the scientific editorial functions. That someone could be a senior editor, an editor-in-chief, one of the coeditors whose office is designated as the headquarters of the journal, or the head of a central secretariat such as in system F.

System E is simply an expansion of system D and carries decentralization just about to the limits of practicality. In system E, the subject matter of the journal is divided into a larger number of subtopics than in system D. Each subtopic is sufficiently narrow in scope to be well within the area of competence of an associate editor. The number of subtopics may range as high as 30 or 35, and the need for coordination will be even greater than for system D. Thus, papers are submitted directly to associate editors, designated by subtopic, who then conduct the complete review process through to rejection or acceptance. However, this system requires that there be a senior editor in charge who is informed of all submissions and rejections, who collects accepted manuscripts for printing, and who can deal with submitted papers that do not fall clearly into one of the designated subtopics.

System F, used by some major publishers, is a variant of system E with a central secretariat in place of a senior editor or editor-in-chief. The secretariat, usually under the charge of a full-time paid manager or managing editor, receives submissions, assigns them to the appropriate editors or associate editors for evaluation, relays the referees' reports and editorial decisions to authors, accepts the manuscripts approved for publication, and is responsible for all subsequent steps in the publication process as well as the business affairs of the journal.

Advantages and Disadvantages

Centralized editorial systems, such as A and B, have the advantages of speed, uniform standards, and control. Postal services in most countries seem to require at least five to six days per mailing these days. Thus, each mailing step in the review process requires at least one working week, and a centralized system keeps the number of mailings to a minimum. Since the editor, in systems A and B, is the only person involved in assessment of manuscripts and referees' reports, a uniform standard is easily achieved. Whether that standard is high or low will depend upon the editor's good judgment, knowledge of the field, and ability to pick good referees. Finally, with all correspondence going directly to and from a single desk, there is superb control. The editor in a centralized system has full information on the status of each manuscript at all times, can make sure that deadlines are met, and runs a minimal risk of losing anything.

The disadvantages of centralized systems (A and B) are the work load that they impose on the editor, the risk of poor selection of referees, and an overdependence on one person. In a centralized system, the evaluation of manuscripts and revisions, selection of referees and assessment of their reports, and correspondence with both authors and referees, all rest with the editor. This work load, which will vary with the size of the journal, may become overwhelming, with the result that editors will either resign or be forced to give up the very research activities that made them desirable editors. The selection of referees is best done by someone who is knowledgeable in the subject matter of the paper. Specialists, searching for referees in their own specialty, can draw on their own knowledge of people in the field, know which journals to use, and are able to assess the contents of the published papers of potential referees in relation to the paper to be reviewed. Some few editors, senior scientists with broad backgrounds and experience, may be able to select referees for the majority of papers in the fields covered by their journals. However, in a fully centralized system such as A, the selection of referees may be less reliable for those papers that are not directly within the editor's specific area of competence. A journal that is operated under a fully centralized system (A or B) runs the risk of becoming a one-person show and may develop an unhealthy overdependence on the editor. With everything flowing over just one desk, nobody except the editor is informed about the journal, and there are few mechanisms for feedback. The editor may become isolated, even insulated, and unaware of current new directions that could affect the journal. To avoid these disadvantages, the editor should maintain an active correspondence with members of the editorial board and should hold frequent (at least once a year) meetings of that board for an exchange of views. A centralized system is totally dependent upon the editor, who should therefore develop a backup in case of emergencies. Ideally, the backup should be an assistant or associate editor, located in the same place as the editorial office and sufficiently well informed to be able to take over in the editor's absence.

The advantages of a decentralized editorial system, such as C, D, E, or F, are distribution of work load and improvement in the review process by having specialists on the editorial board deal with papers in their fields. In these systems, associate editors or coeditors are selected to represent the significant areas of subspecialization within the editorial scope of the journal. A retrospective analysis of the number of papers submitted in each subspecialty provides the key to distribution. Thus, a journal that receives say 500 submissions per annum, distributed more or less evenly over 10 subspecialties, might have 10 associate editors each of whom would deal with 50 papers a year, or about one per week. Clearly, the work load could be reduced further by the appointment of more associate editors, more than one per subspecialty if necessary. Since the associate editors, in decentralized editorial systems, deal with papers in their own specialties, the selection of referees is both easier (less time-consuming) and more reliable than it would be for a single editor trying to cover all areas of subspecialization.

The disadvantages of decentralized editorial systems are a slower response in the review process, greater variability in standards, and weaker control than in centralized systems. For example, in system C the editor receives submitted manuscripts and mails them to appropriate associate editors who then mail them to referees. In the reverse phase, the referees mail reports to the associate editors who add their comments and then mail the reports, comments, and manuscripts back to the editor. The editor then makes a final assessment and mails the manuscript, with a decision letter, back to the author. Starting from the editor, there are five mailing steps to complete the review process in system C as compared with three mailing steps in system A. The number of mailings of manuscripts may be kept to three in systems D and E, in which submissions are made directly to coeditors or associate editors. However, manuscripts that are borderline with respect to subtopic may get passed from one associate editor to another, and extra mailings of forms and copies of correspondence are required to coordinate these systems. By and large, systems C, D, E, and F, simply because they involve more people, will tend to be slower than systems A and B.

The larger number of people involved in the decentralized systems can also introduce variability in standards. The various coeditors or associate editors cannot be expected to be identical in the promptness with which they deal with journal business, in the care they devote to selection of referees, in their concepts of what constitutes an acceptable paper, or in their styles of correspondence with authors. This variability can be reduced, but not eliminated, by meetings that can be used to indoctrinate new associate editors and to develop a consensus on important issues.

The more people involved in an editorial system, the greater the problem of control. Possibly the worst difficulty is having things fall between the cracks; i.e., everyone involved assumes that someone else is dealing with a certain item when, in fact, nobody is and that item just sits there, forgotten. With papers flying hither and yon in a decentralized editorial system, there is an increased risk of actually losing something. For example, an editor assigns a manuscript to an associate editor, but several months later, following phone calls from the irate authors, nobody can track down the manuscript. Was it sent out at all or did it get put in the wrong pile in the editor's office? Was it sent inadvertently to another associate editor? Was it sent to the correct associate editor but then misplaced in that office? Was it sent to referees with no record noted? Did it get lost in the mails? Mistakes of this kind can and do happen in any editorial system; after all, humans are fallible and as subject to Murphy's Laws as anything else. However, it is clearly more difficult to maintain the necessary checks and controls in a large, decentralized editorial system than in one that is small and centralized. System F is designed to overcome some of the problems of control by having a central secretariat as the contact point for everything having to do with the journal. The advantages of this system are:

- Scientific editors and associate editors are relieved of the administrative burdens of controlling and recording the flow of paper.
- There is one permanent address for the journal, and it does not change when editors or associate editors rotate.
- All information about the journal is in one place, thus facilitating control and record keeping.
- The system can handle a very large flow of manuscripts.

The disadvantages lie in having non-scientists rather than well-known scientific editors as contact points for the journal; a central secretariat is also expensive and affordable only by major professional publishers.

Which System for Which Journal

It is impossible to give specific guidelines as to which editorial system might be best suited to a journal because there are too many variables, some of which are unknown and unpredictable. However, the major factors that should be considered in choosing an editorial system are the styles and personalities of editors and associate editors, the size of the journal, and the degree to which the journal is specialized.

Editors have different personalities and should be free to select the editorial system with which they feel most comfortable. It is counterproductive to force editors to work within systems in which they are unhappy. Thus, an editor who likes to maintain a "hands-on" operation, with a minimum of delegation, is not necessarily autocratic, but will be happiest with centralized systems such as A or B (Table 1). System C, in which the editor sees everything and does all correspondence with authors, could be a viable alternative for that kind of personality when a journal is simply too large for one person to handle. On the other hand, editors who are good managers, confident and easy in the delegation of responsibilities, will be happiest with one of the decentralized systems, C, D, or E, regardless of the size of the journal. Editors who prefer a minimum of administrative detail will work best under system F.

The size of a journal determines the editorial work load, and there comes a point where that work load must be distributed for the sake of efficiency. A number of manuscripts up to, say, 150 per annum is probably well within the capacity of a single editor who is reasonably organized. There may even be some editors who handle 300 or more manuscripts per annum with no signs of distress. However, many editors faced with that kind of a work load will want and need the assistance provided by a decentralized editorial system. With very large journals the administrative burden requires the full-time attention provided in system F.

Because referees are best selected by someone who is knowledgeable in the subject matter of the paper, it is obvious that general journals require a greater diversity of editorial expertise than specialized journals. There are probably some journals with a sufficiently narrow editorial scope that all papers submitted fall well within the range of expertise of an individual. If those journals are also small, they can be edited efficiently and well by one person under the fully centralized system A. In journals with a broader editorial scope, editors should supplement their own expertise by the use of associate editors to cover specific subtopics.

Broad general journals require editorial representation for each significant subtopic of the field, either by large groups of associate editors (system C or E) or by subdivision into major sections, each with a co-editor and associate editors (system D).

Meetings

Editors should try to convene meetings of their editorial boards at least annually as a matter of good management. Annual meetings serve a number of purposes: everyone can be updated on the status of the journal, the editorial system can be reviewed and suggestions for improvement considered, new associate editors can be introduced to the system, instructions to authors can be reviewed for improvement, and matters of policy can be discussed. Finally, and by no means least, an annual meeting lets the members of the editorial board get to know one another, stimulates their interest in the journal, and provides for an exchange of views that is often useful and enlightening. Editorial boards can meet most conveniently at society conferences or international congresses which most of the members would attend anyway, thereby sparing the journal the costs of travel.

Journal Records

The detailed records of a journal operation may be an uninspiring bunch of numbers, but an analysis of the data can often be revealing in a useful sort of way. A simple annual record of total submissions shows whether a journal is growing, shrinking, or standing still. The trend in submissions, taken in relation to activity in a field, can indicate whether or not a journal is attracting its fair share of papers. The results may sound an early warning that some positive action is required, and the editorial board can be probed for suggestions. In non-specialized journals, an analysis of submissions by subtopics can sometimes indicate a change in emphasis or direction in a field. If one subtopic is shrinking, perhaps it should be dropped and a new one created to provide a common home for a number of the papers submitted to a growing miscellaneous category. Analysis of the source of papers might reveal trouble spots that require attention. For example, submissions from a certain laboratory or department have suddenly dropped off. Did the principal authors move, die, or switch to a new journal that is more appropriate, or were they unintentionally offended by the treatment that their last paper received? Conscientious editors will want to follow up sudden changes in sources of papers; not that authors should be begged or mollycoddled, but, if they are being alienated, editors should find out why. Another useful statistic is the average time required for publication of papers. Scientists put a high priority on speed of publication, and editors should make every effort to keep their journals competitive with others in the same field. The interval between dates of submission and dates when manuscripts are returned to authors for revision is a measure of the efficiency of the review process. Obviously, there is a practical lower limit of some two to three weeks, and that assumes ideal performance by the postal service and the referees. However, there ought also to be an upper limit or target, and a period of about seven weeks is reasonable for journals in most fields. (Papers in mathematics take much longer to referee, frequently several months.) Some slippage has to be accepted to allow for natural disasters, such as postal strikes, and for referees who decline or are unavailable. However, if any of the associate editors on an editorial board show consistently longer times than the others for the review process, it is an indication that they either neglect their mail or fail to send reminders to referees. Editorial boards will also be interested in the time required to copy edit and print accepted papers. Any slowdown in either of those processes should be a matter of concern and cause for inquiry by the editor. Finally, editorial boards should be interested in financial statements that include costs, subscription prices, and circulation. Even if the editorial boards are not responsible for the finances, the members should know what it costs to produce the journal, its circulation, and its sources of income.

Review of Editorial System

The annual meeting of the editorial board is the ideal time to review the editorial system used by the journal. There are, first, the general aspects such as: Is the present system appropriate for the journal, or should it be more centralized or decentralized? Are there enough, too many, too few, associate editors; do they represent adequately the subtopics of the journal; and is the work load distributed equitably? Then there are the more detailed aspects such as who does what and when. In decentralized editorial systems, there should be manuals of operating procedures and guidelines for each person involved. These manuals should contain the following items:

- a description of the editorial scope of the journal
- descriptions of the categories of contributions — articles, reviews, notes, communications, letters, etc.
- a statement, reflecting the consensus of the editorial board, about criteria for acceptance

- guidelines for finding the best referees
- an outline of administrative procedures which may include a manual for secretaries
- a sample of each form, form letter, or card that is used in the editorial system

Changes in any of these items can be discussed and implemented at the annual meeting of the editorial board; the goal should always be to improve the system in terms of efficiency and standards.

Indoctrination of New Members

The occasion of the annual meeting provides an opportunity to impress upon new members of the editorial board the importance of referee selection. The more experienced members of the board can explain how they find referees from references in submitted papers, and which journals, indexes, or directories are most useful as sources of names and addresses. A discussion of any problem papers or ethical issues that may have arisen during the preceding year, and how they were handled, would be profitable for new members of the board. Finally, the editor may want to provide new members of the board with a few examples of correspondence with authors as a guide to style. It is recognized that letters to authors about revision of their papers must be individualistic because each paper is presumably unique. However, samples of previous letters can show new members of the editorial board how to vary the degree of emphasis in their directions to authors about revisions.

Review of Instructions to Authors

Instructions to Authors contain a description of the editorial scope of a journal and define how manuscripts should be prepared for submission. Instructions are necessary if a journal is to maintain any sort of format and standards of presentation. From time to time, the Instructions to Authors need to be amended to reflect a change in editorial scope, to improve the efficiency of copy editing, to incorporate new standards of nomenclature or to accommodate new printing methods. A review of Instructions is therefore an appropriate item for a meeting of the editorial board, but it should be done in consultation with a managing editor or whoever is responsible for copy editing and printing because those latter two functions are the ones most affected. Details on the development and use of Instructions to Authors are given in Chapter 8.

Policy Items

Finally, there are several policy items that should be reviewed by editorial boards, preferably in consultation with the publisher, whose interests will be involved in a number of them.

What is the role of the journal? Is it strictly an archival journal of primary research, or is it also expected to carry society news, annual addresses by society presidents, obituaries, and announcements of meetings? If the journal is expected to have a news function, then how are those items collected and edited, and what guidelines for acceptance should be used? For example, if announcements of meetings are published in the journal, should these be limited to local, national, or international? If obituaries are carried, should they be simply name listings or should they include a description of the careers of the deceased?

If the journal does not have a news function, then what about the publication of reviews? A number of journals of primary research have found that publication of the occasional review, usually as a lead article, has been received very well. If reviews are published, should they be critical or comprehensive, or both? Are autobiographical reviews by eminent scientists acceptable? How many reviews should be published, one each issue or only one or two each year? How will reviews be selected? Will they be invited? How will they be refereed, if at all? Should the journal publish annually the lecture given by the winner of the society's most prestigious award?

Does the journal carry advertising, and if not, should it? Some people feel that commercial advertising detracts from the scholarly appearance of a primary journal. However, in some fields advertising can make a journal more attractive to readers by informing them of new products or equipment. If advertising is carried, is it handled by the publisher or by an agent? Is the agent doing a good job, or should a change be contemplated? Is there sufficient control over unethical advertising, particularly in the biomedical field?

What about book reviews? Some editors feel that book reviews are a useful service to readers, others find them redundant — who needs 15 different reviews of the same book? If book reviews are included, who selects the books and what criteria are used?

Should the journal publish special issues, and if so, should these be published as part of the regular run or as supplements? Some journals have found that special issues, with groups of papers devoted to one topic, have become best sellers and attracted favorable attention to the journal; other journals avoid them like the plague because of uneven quality of the papers and because they disrupt the normal schedule. Such groups of papers often arise from a conference or symposium, and certainly a special issue in a regular journal is a much better forum for publication than a separate, isolated volume of Proceedings that becomes lost and almost irretrievable. However, a cautionary note is necessary. Not all papers presented at a conference are ready for publication. Therefore, journals that agree to publish special issues should require that all papers be subject to the normal process of review.

Another form of special issue is the Festschrift, an issue that is prepared specifically to honor an eminent scientist, in the field covered by the journal, on the occasion of that individual's birthday, retirement, or winning of a prestigious award. Again, a cautionary note is in order. Editorial boards should develop clear policies that are applied uniformly and consistently in dealing with requests for special issues in honor of individuals. These requests arise from friends and colleagues who will have strong, personal, and subjective feelings about the person being honored. They will therefore press editors for the most elaborate treatment possible, and editors who are inconsistent run the risk of offending sensitivities. The following questions might be considered: Should dedicatory issues be published at all? Should the dedicated papers be reviewed in the normal way? Who merits a dedicated issue and what criteria should be used? Should there be a dedication, and if so, how long should it be—one short paragraph, one page, or a full-blown biography? Should a photograph of the individual being honored be included?

It is not possible to make specific recommendations on the foregoing policy items because different editorial boards and publishers will have their own points of view and will have to decide what is best for their journal. The description of these various items is included here simply to show the kinds of things that editorial boards may want and need to thrash out at annual meetings.

Chapter 4

The Review Process

Are Referees Really Necessary?

There are certain scientists who regard all referees and editors as biased adversaries whose objectives are solely to reject, delay, or scoop all papers submitted to them. It is unlikely that those attitudes can be changed by any amount of argument; editors accept that as a fact of life. Indeed, Goudsmit[12] has suggested that a touch of paranoia may be a necessary condition of a successful research worker. Authors with those inclinations have proposed, from time to time, that the scientific literature could function quite well, and maybe even better, without referees. The argument goes that new ideas and information could be published without delay and that science would therefore advance more quickly. Editors would accept anything that was readable, and the scientific community at large would sort out the wheat from the chaff after publication. The problem is that such a system would yield too much chaff. The proponents of unrestricted access to publication have failed to distinguish between the formal and informal systems of communication in science. They have the well-meant but misguided concept that the success of the free and easy exchanges in generating new ideas at meetings can be carried over into the literature. However, this concept overlooks the absolute necessity, for scientific progress, that somehow, somewhere, there has to be a reliable record. Unrestricted publication would turn the scientific literature into a meaningless jumble in which nobody could find the significant papers. The result would be total chaos and would apply to form as well as substance. In the context of urging the use of less stilted language in scientific papers, Price[23] noted

> that the scientist is more frequently than not passionate, biased, illogical, resistant to proof and to change, and beset by other similar human failings. It is, today, better to let this show and be understood than to pretend that it is not there.

43

"THAT'S IT? THAT'S PEER REVIEW?"

As valid as this statement may be, it is more applicable to the informal than to the formal system of communication. Surely, the unconstrained use of intemperate, subjective language, reflecting biases and prejudices, would make it even more difficult to assess the content of a paper. Scientific arguments need to be presented objectively if they are to be debated rationally. One function of the refereeing system that is often overlooked is its indirect influence on the initial preparation of a paper. Established scientists write their papers with a critical sense that anticipates referees' questions. Without this subtle pressure in the background, there can be little doubt that the quality of presentation would deteriorate along with the content.

The constraints on scientific publication are not new; they began with the very first journals. Publication was limited to those reports that were read at a meeting of a learned society or submitted by a member. As science grew, those limits became too narrow to accommodate all the worthwhile papers. Editors then began to consult colleagues and members of their societies about the interest and veracity of papers that were submitted. Thus began the system of referees, anonymous experts who are called upon to advise editors on the suitability of papers for publication. The role of referees is more contentious than any other aspect of scientific publication, but alternatives are not easy to find. If there is agreement that some form of monitoring is essential, and the case for that seems overwhelming, then who is capable of doing it other than experts in the subject matter of the paper?

Does the System Work?

The answer to this question is a resounding, Yes! All editors, and most authors, will affirm that there is hardly a paper published that has not been improved, often substantially, by the revisions suggested by referees. The same editors will state that examples of intentional delay, biased reports, or unethical behavior are extremely rare. So how do we account for the spate of criticism in recent years, and what, if anything, is wrong?

The Different Processes of Peer Review

A good deal of the criticism is misdirected because it fails to distinguish between the different kinds of peer review: evaluation of papers for publication, assessment of grant proposals, and recommendations for positions or tenure. Referees of scientific papers are really being asked just one question: Is this paper suitable for publication? The questions asked by grants panels and tenure committees are quite different. A review of a paper is retrospective whereas the review of a grant proposal is prospective, and there is a world of difference. Both grants and tenure are influenced by the applicant's track record in science, but that record is irrelevant to the review of a specific paper. Considerations of available facilities, budgets, and personal characteristics are important for grants and tenure but do not enter into the review of a paper. It should not be surprising that reviewers' assessments differ when they are asked to consider such a range of topics: soundness or promise of proposed research, budgets, significance of past research, facilities available, personal characteristics, teaching ability. Furthermore, the awards of grants or tenure are concerned with competition for money and therefore induce, however subliminally, a complex mixture of motivations in both applicants and reviewers. This is not the place to present a critique of the review processes for grants or tenure. However, it is important to recognize that those review processes differ, in both complexity and objectives, from the review of scientific papers for publication. General criticisms of "peer review" often do not take into account those distinctions.

There is one circumstance in which the different peer review systems affect each other, and that is in the pressures to publish or perish. The background noise in the scientific literature could be reduced greatly if grants panels and tenure committees gave more attention to reading rather than counting. What is needed is a clear demonstration, to applicants for grants or tenure, that a few good, substantial papers would be more highly regarded than a long list of fragmented or duplicate publications in the form of letters, communications, conference proceedings, etc.

Two Kinds of Consensus

There are times when referees will disagree about the suitability of a paper for publication, and so will journal editors. Not infrequently a paper that is rejected by one journal will be accepted by another. Such disagreements do not indicate a failure or breakdown of the referee system; rather, they arise from differing views of two kinds of consensus.

First, there is the consensus that develops within an editorial board on what constitutes an acceptable paper *for that journal*. That consensus may well differ from one journal to another because it is usually based upon some perceived degree of significance. Of course, the required degree of significance cannot be spelled out because it is a judgment call, difficult to define in general terms. Thus, acceptance of a paper by one journal after rejection by another may simply reflect a difference in the perception of an acceptable degree of significance by the two editorial boards. As most editors try to maintain or raise the standards of their journals, some variation in acceptable degree of significance may be expected.

The other consensus that enters into the evaluation of a scientific paper is the one that develops within a field of research. In some areas of science the fundamental algorithms and paradigms are well established, tested, verified, and accepted. There is a consensus about the state of knowledge in those areas that has two consequences in the realm of scientific publications. The first is that referees of a paper in a well-established field will seldom disagree on matters of substance. The second consequence is that any paper that ventures outside of the consensus of the field will be questioned severely. In such instances referees and editors may be accused of censorial behavior, stifling imagination and brilliant ideas. However, a moment's reflection reveals the soundness of questioning anything that is much at variance with a consensus that has been built up by a long record of tested, verified work. That questioning protects the scientific record from frauds, cranks, and even insanity. It is fully appropriate to require the strictest tests of confirmability and repeatability for any work that purports to upset the solidly established base of knowledge.

There has been at least one attempt to provide an outlet for speculative ideas that go beyond the consensus of knowledge. The journal *Speculations in Science and Technology* was started in 1978 with the declared intention of publishing speculative papers in the hard sciences. The policy of that journal has evolved with experience as explained in editorials, portions of which are reprinted annually. The journal is intended to be a forum for ideas that may not be supported by the current consensus, but the exposure of which may help to clarify the limits and basic assumptions of our accepted ways of thinking. Even with this liberal policy, the editor has had trouble with authors who failed to show how their idea related to the present state of the problem and who ignored, or showed no knowledge of, extensive previous work on the same topic.[16]

Despite such difficulties, *Speculations in Science and Technology* may serve as a useful safety outlet for those rare good papers so far ahead of their time that they are rejected by journals that keep to the normal constraints of scientific publishing.

Legitimate Controversies

In contrast to those fields of research that have a solid consensus of knowledge, there are areas where a consensus has not yet developed. Such areas may be at the research frontier of a well-established field, e.g., some aspects of theoretical chemistry, or may encompass a whole subject, e.g., cosmology. Legitimate controversy and competition are the main characteristics of those areas of research and are in fact how they advance. The controversies usually involve disagreement about the meaning of results when there is not enough information to decide clearly between two or even several plausible interpretations. Eventually, further experimentation and research will resolve the issues. Until that happens, however, science is best served by the publication of all sides of the controversy, provided that each paper contributes some new information or insight. Some controversies linger on for a long time; others are settled reasonably quickly. For example, the concept of continental drift was debated for some 50

"IT STARTED WITH A SIMPLE CASE OF PEER-REVIEW."

years whereas the issue of polywater (or anomalous water) was settled within 10 years.

Editors should be pleased to attract to their journals papers in new, exciting areas of science that are advancing rapidly. However, editors must recognize that such papers may be controversial and require careful treatment. Several options are open: accept and publish any reasonable paper (1) without comment, (2) with comments by a referee or a known adherent of another side of the controversy, (3) with simultaneous or subsequent publication of comments by a number of critics representing all sides of the controversy, and (4) with inclusion of rebuttals by authors for either of the latter two options.

The choice from among those options is largely a matter of editorial preference. Most journals in the natural sciences tend to adopt option (1) with occasional ventures into option (2). Their choice is based upon the following rationale: "Here is a paper in a controversial area. The referees say that it is reasonable and contributes new information. We will therefore publish it and let the adherents of other sides of the controversy counter it with papers of their own. That is the way controversies are resolved."

Every now and then the interpretation or conclusions in a paper may be disputed by a referee. In such cases, if editors can establish that both viewpoints have some validity, the situation can be defused by inviting the referee to submit a comment or discussion to be published together with the original paper and with the authors' knowledge. That leads to option (2) above and provides for the airing of both sides of the issue simultaneously.

Open Review

Option (3) above constitutes open review in which a paper is sent simultaneously to several referees who are invited to prepare comments for simultaneous publication with the paper and rebuttal by the authors. This system, used in the journal *The Behavioral and Brain Sciences*, is intended to alleviate conservatism and encourage creative disagreement.[14] Open review may well be beneficial in the social sciences, where there is not a strong consensus of knowledge. This lack of consensus is, incidentally, fully understandable in view of the large number of variables and the difficulty of controlling them in social science experiments. Under such circumstances even the proper design of experiments is debatable, and open discussion may throw up a whole variety of useful ideas. However, the use of open review is of less obvious value in fields of research where there is a well-established consensus of knowledge. There would be little point in publishing referees' reports on papers in the natural sciences because the vast majority of them do not contribute anything new in the way of ideas or concepts; they simply tell authors how to do it better. A few reports will point out major or minor errors that authors overlooked, and publication of those reports would be needlessly embarrassing to both authors and referees. Similarly, nothing would be accomplished by seeking or publishing reports from 10 or 12 referees because they would be repetitive. Finally, the publication of referees' reports would abrogate the principles of anonymity and confidentiality that are crucial to the effective operation of a refereeing system.

Anonymity of Referees

Despite the proven usefulness of anonymous refereeing, this issue remains controversial. Referees are sometimes accused of using the cloak of anonymity to conceal personal biases, unethical behavior, and incompetence. In fact, those accusations generally do not reflect what is happening. Scientists are both authors and referees, and the balance between these two functions suppresses unreasonable behavior. A referee's viewpoint may well differ from an author's in controversial areas, but that is a scientific rather than a personal bias. Referees very seldom engage in unethical behavior such as stealing ideas or delaying manuscripts deliberately, and referees who feel that a paper is beyond their competence to review will return it to the editor with a note to that effect. The foregoing statements would be affirmed by the editors of most journals and could be proved by examination of their files. Every editor can cite instances in which authors have complained about the choice of referees, accusing them of being incompetent or even idiotic, when in fact the referees in question were suggested as experts in the field by the authors themselves. Would those same referees have given the same reports if they knew that their identities would be revealed to the authors? Probably not, and that is the prime reason for maintaining anonymity of referees; editors need to know what referees really think. Given the vagaries and complexities of the human condition in modern science, referees must be free from the possibilities of repercussions if they are to be expected to report with candor and honesty. Young scientists would fear to criticize papers whose authors might subsequently be reviewing their grant proposals or applications for tenure. Many scientists do not want to jeopardize friendships or collaborations by the uncertain reception that might be given to an identified, critical review. Finally, in the increasingly litigious atmosphere of today there is even the possibility of referees being sued. Remote as this latter possibility may be, the contemplation of the hassle, expense, and time involved in a court proceeding is a major deterrent for many referees. Perhaps these various repercussions would never happen. Perhaps, like the personal biases and incompetence of referees, they are imaginary. The difference is that editors *know* that most accusations leveled at referees are false; they do not know that fears of repercussions on the part of identified referees are imaginary. On the contrary, editors believe that those fears are very real, at least in the minds of referees, and that is what matters.

A point often overlooked by critics of the anonymous refereeing system is that referees advise editors, not authors, and editors may or may not follow that advice. True, editors may use referees' reports in whole or in part to help authors improve their papers or to support a decision of rejection. However, good editors will make clear what points in the reports they consider important and how they reached their decision. Authors therefore deal directly with editors who make the decisions to publish or reject and who are not anonymous. To make sound decisions in this day of complex, highly specialized science, editors need unrestricted access to the best advice available; those conditions are best met by the anonymous refereeing system.

Although editors should make every effort to preserve anonymous refereeing, there will be some referees who want to identify themselves to authors. It may be that the referees want to discuss further the implications of a paper they have just reviewed, or they may want to propose a collaboration with the authors. Editors have nothing to lose by acceding to such legitimate requests. The important point is that referees' requests to be identified be made for some specific purpose; otherwise, anonymity should be both encouraged and guaranteed.

Blind Review

Another variation of the refereeing system is to remove the authors' names from the paper before sending it out for review. This system is known as blind review and has been advocated as a way to reduce supposed biases induced by authors' reputations or lack of them. However, experience has shown that blind review cannot conceal authors' identities and is simply not worth the trouble. What happens is that authors who are well established are easily identified by the work reported in their papers or by such phrases as, "Earlier work from this laboratory has shown . . . ," and "We reported previously that. . . ." Authors who have not published previously in a field would not be known to referees anyway, so the presence of their names makes no difference. It is like the story of the chess master who was invited by a passing visitor to play an informal game. The master removed his own Queen's rook before the game started whereupon the startled visitor exclaimed, "You've never seen me before, how do you know that you can give me the odds of a rook?" The master replied, "If I knew you, I would not have to give you odds." The chess master knew personally all of those few chess players in the world who were of his calibre. So it is in each specialized area of science, and that is why anonymity of authorship is either impossible or pointless.

Summing Up the Review Process

Without some form of monitoring its contents, the scientific literature would become unreliable, confused, and unreadable. The most effective monitoring system is the use of expert referees who provide editors with advice about the suitability of papers for publication. Criticism of the peer review system in general is often misdirected as far as journals are concerned because it fails to distinguish between reviews of papers, reviews of grant applications, and recommendations for positions or tenure. Differences in the results of editorial reviews may arise in two ways: (1) editorial boards of different journals may vary in their perception of the degree of significance required to make a paper acceptable, and (2) referees may disagree about papers on controversial subjects or in fields where a consensus of knowledge has not developed. Legitimate controversies are best resolved by providing for appropriate publication of all points of view. Anonymity of referees is essential to the proper functioning of the review process in journals.

There is nothing wrong with the basic principles of the existing review system in journals. If, as some claim, there is some deterioration of quality because of the pressures of increased scientific activity and the publish or perish syndrome, the answer is surely not to throw out a sound system but to make it work better. Journal editors are surely the key people who can introduce improvements. It is they who establish the acceptable degrees of significance for their journals, make the decision to accept or reject papers, and, most importantly, select the referees for each paper.

Chapter 5

Referees

Importance of Selection

The choice of referees for a paper is the single, most important decision that editors make in dealing with submitted manuscripts. The correct selection of referees makes all subsequent steps relatively untroubled and straightforward. This is not to suggest that editors delegate their decision-making responsibilities to referees. Referees advise editors but that advice must then be applied intelligently, and that is the editor's function. However, if the advice is appropriate and properly presented, the editorial decisions become very easy to make.

Because the selection of referees is so important, it follows that it should be done carefully; this is not the place to take shortcuts or to make snap judgments. The time spent in finding the right referees for a paper is well worthwhile and will save time later on. These comments apply to anyone who is selecting referees. If, as often happens, editors delegate the selection of some referees to coeditors, subeditors, associate editors, or members of editorial boards, each of these individuals should be carefully coached on how to find the right referees.

How Many

The first concern is how many referees are needed for each paper. Probably the most common practice is to send a manuscript to two referees simultaneously. The merits of this system are that it usually provides at least one solid report, that the two reports can be checked against each other, and that one referee may cover points that the other missed. If two referees per paper is a good system, would not four, six, or eight be even better? True, the chances of obtaining reports that agree would be greatly increased with a larger number of referees. Editors could then

53

rely on the majority opinion and discount one or two reports that might disagree. However, this advantage would be offset by the added demands on referees; by the added expense in time, postage, and extra copies of manuscripts; and by the fact that most of the multiple reports would be repetitive. Thus, the assignment of two referees per manuscript provides for enough checks without overloading the system unnecessarily. While choosing two referees for a paper, editors may want to select a third for possible use later should one of the first two not respond, or in case a third opinion is needed to sort out disagreeing reports.

How to Find

How can editors find the right referees for a paper? An editor who is active in research will know personally the hundred or so scientists in the invisible college of that particular specialty. When a paper in that specialty is submitted, the editor can list, by memory, the names and addresses of half a dozen good referees. Of course, the same applies to coeditors and associate editors when they are selecting referees. Thus, a journal's best source for selection of referees is an editorial board of knowledgeable scientists whose areas of expertise reflect the major topics within the editorial scope of the journal. An editor's personal knowledge of referees who provide good and reliable critiques is the journal's most valuable resource.

The next best source is probably the journal's record of referees because the interests and performance of people listed in that record are known quantities, even though editors may not know the referees personally. The risks in editors limiting themselves to a list of referees are that the list becomes both outdated and inbred. Scientists change their research interests and directions with surprising frequency but may be reluctant to admit that they are not current in a former specialty. A list of referees may contain, say, 6 to 12 names of scientists in a particular specialty or subspecialty. With no change or expansion in the list these 6 or 12 experts will end up refereeing each other's papers in perpetuity. Thus, if editors want reports that reflect the current scene and fresh viewpoints, they should not rely too heavily on a stable of referees; remember, a stable does not contain many winners.

A refinement on a list of referees is to match referees with papers by computer, using key word profiles. In this process, referees are asked to describe their expertise in key words or to tick off key words on a list. These key word profiles of expertise are then matched with key word profiles of submitted papers drawn up by either the authors or the editors. The success of referee selection by computer matching depends very much upon the skill in key word profiling and the extent to which it can penetrate a specialty or subspecialty. Editors who want to venture into this process should therefore obtain the assistance of the best experts in information retrieval. Most good science libraries will have at least one such expert on staff.

Another good source of referees is the list of references in submitted manuscripts. The key references are easy to spot during a brief scan of the paper, and the authors of the cited papers are good potential referees, particularly if the cited papers are reasonably current, say within the past two or three years. If the references are not to current papers, the authors can be looked up in author indexes to see if they have published more recently and if their interests still fit the subject matter of the submitted manuscript.

Some journals invite authors to suggest possible referees when submitting a manuscript with, of course, no guarantee that they will be used. Editors who use that system will usually select one referee from the authors' suggestions and another of their own choosing. It is probably just as important to ask authors if there are people to whom their paper should not be sent and to respect those wishes; there is no point in starting a war unnecessarily.

There will be times when editors and associate editors receive papers that are not directly in their areas of expertise and when no obvious leads to referees are provided by either the authors or the reference list. When that happens, editors have no choice but to turn to the literature to find people who are working as closely as possible to the subject matter of the submitted paper. Names can be obtained from various directories of graduate research or from society compilations of research in a certain field. Recent subject indexes of journals in the same field can be searched for papers related to the topic of the manuscript; abstract journals can be used in the same way. Under these circumstances, when referees are not known to the editors either personally or by past performance, editors may want to look up and at least scan one or two of the potential referees' papers. This will give some idea of the standards and current interests of the referees and their relevance to the submitted paper. In fact, editors would probably benefit from more frequent forays into the literature to find referees. Of course, editors are busy people and understandably reluctant to spend the necessary 10 to 15 minutes per paper in the library looking up referees for *every* paper. However, the effort will prove worthwhile for those papers for which there are no obvious referees or where editors have any doubts at all about referees selected from other sources.

Guidance

Journals, and their editors, vary considerably in the amount and form of guidance that they give to referees. The variation ranges all the way from no guidance at all to a highly detailed package that contains, as separate items:

- a guide to referees
- a checklist for referees
- a form for the referee's report to the editor
- a form for the referee's comments for the authors

Some journals send out manuscripts for review with a very simple covering letter, "Could you please review the enclosed paper as to its suitability for publication in this journal. We would be grateful if you could return the manuscript with your report within two weeks." Attached to this letter there is an uncomplicated referee's report form, probably with spaces to check off (acceptable as it stands ____, acceptable with minor revision ____, acceptable with major revision ____, not acceptable ____) and a blank page for the referee's comments. Figure 2 shows an example of this kind of form. This system certainly has the advantage of simplicity and must work reasonably well because it is used by several prestigious journals. Editors of those journals assume that referees are familiar with the traditional constraints that apply to scientific publication. More to the point, perhaps, by not leading the referees or forcing them to respond in a set format, the editors hope to get reports that are more free and pertinent. In effect, the editors are asking simply, "What do you think of this paper?" and expect that a reply, "I think it is great because . . ." or "I think it stinks because . . . ," will be more useful than ticked off answers to a set of predetermined questions. With so little guidance, how do new referees know what is expected of them? The answer is, of course, that referees are also authors and have seen how their own papers were reviewed.

Some editors feel that referees need to be reminded of certain points that are considered to be important to their journals. Those points may involve questions about the appropriateness of the journal, the possibility of reducing detail, and the use of correct nomenclature, e.g., Figure 3.

Other journals issue a more detailed form for referees' reports with questions about the scientific quality, references, prior publication, tables and figures, and length, and by this means provide the criteria by which the papers should be judged. For examples of this kind of report form, see Figures 4 and 5.

Still other journals send out guidelines to referees that are separate from the referees' forms. Again, there is considerable variation in the contents of these guidelines; some (e.g., Figure 6) deal primarily with definitions of editorial scope and the need for conciseness, others (e.g., Figures 7 and 8) spell out the traditional constraints on scientific publication and

IF YOU CANNOT RETURN THIS MANUSCRIPT WITH COMMENTS IN **TWO WEEKS**,
PLEASE RETURN IMMEDIATELY WITHOUT COMMENTS

The Journal of **Organic Chemistry**

Reviewer _____ Ms # _____ Date _____

Author(s):

Title:

Comments:

In my opinion, this manuscript should be published:

| without | with minor | with major | |
| □ revision | □ revision | □ revision | □ Not at all |

Please sign original only and return
with one anonymous carbon.   **Signature** _____

Figure 2

IF YOU CANNOT RETURN THIS MANUSCRIPT WITH COMMENTS IN TWO WEEKS, PLEASE RETURN IT IMMEDIATELY WITHOUT COMMENTS

_____

JOURNAL OF THE AMERICAN CHEMICAL SOCIETY

PLEASE RETURN SIGNED ORIGINAL AND ONE ANONYMOUS CARBON

COMMENTS OF REFEREE _____

AUTHOR:

TITLE:

Would the article be suitable for publication:

1. without change? _____ 2. after minor revision? _____ 3. after major revision? _____
Is JACS the best medium for publishing this article? _____ If not, what other Journal would you suggest? _____ Please be as specific as possible if revision by the author is recommended. Indicate specifically whether descriptions of method, tables of data, etc. should be reduced or eliminated with the understanding that they would be available to the specialist in the form of microfilm or in some other way. Does the nomenclature used conform with accepted practice as exemplified by Chemical Abstracts and the recommendations of the Committee on Nomenclature, Punctuation and Spelling of the Society? Are hazardous procedures clearly identified as such?

COMMENTS:

PLEASE SIGN THE ORIGINAL ONLY _____

Figure 3

Manuscript No._____

REPORT OF REVIEWER - JOURNAL OF APPLIED PHYSICS

Please send your recommendations and comments and return the manuscript within two weeks to the Editor, Journal of Applied Physics, Argonne National Laboratory, Box 296, Argonne, Illinois 60439. If you know at the outset that you will not be able to review this manuscript within the next three weeks, please return it immediately; suggest another reviewer if possible. Please return two copies of this report, preferably typed. Extended comments and suggestions on extra sheets are helpful to both the authors and editors and are most welcome. When possible, these should be typed in duplicate with one copy so prepared that it can be sent directly to the author without revealing your identity or your affiliation. Questions which may be helpful to you as a check list as you review the paper appear below. If you prefer not to use them, simply write us a letter.

I. CHECK LIST:

1. Is the paper of good scientific quality, free from errors, misconceptions or ambiguities, and does it contain sufficient new results, new applications or new theoretical developments of interest to physicists to warrant its publication in the Journal of Applied Physics? Please indicate on a separate sheet or in pencil on the margins of the manuscript any points which are objectionable or which need attention.

2. Is the Journal of Applied Physics the most appropriate journal? If not, can you suggest a more suitable one?

3. Is the manuscript a clear, concise, reasonably self-contained presentation of the material, giving adequate reference to related work? Is the English satisfactory? Please indicate needed changes in pencil on the margins of the manuscript if you wish.

4. Are the tables and figures clear and relevant and are the captions adequate? Are there either too many or too few?

5. Does the paper make effective use of Journal space, or are parts unnecessary, unimportant, or subject to condensation? If so, which? Write in pencil on the margins of the manuscript if you wish.

6. Is the title appropriate and is the abstract adequate for verbatim reproduction in abstract journals?

7. Please suggest additional or alternate reviewers.

8. Remarks: (Use a separate sheet if you wish.)

II. RECOMMENDATION:

	(As is	☐
1. Publish	(With revision optional	☐
	(With revision required	☐
2. Reject outright	☐
3. Reject, but recommend referral to another journal (See I-2)	☐

050575

Figure 4

WATER RESOURCES RESEARCH
Reviewer's Appraisal Form

Author: Sent:

Title: Returned:

The editorial board would appreciate it if you would review this paper and return your completed review <u>within</u> <u>three</u> <u>weeks</u>. If you cannot meet this schedule, please return the paper immediately.

The following questions and comments are provided as a guideline for reviewers. Please provide detailed comments on a separate sheet; your unsigned review will be forwarded to the author(s). Please avoid phrasing that might generate antagonism. Any confidential remarks for the editors should be included in a cover letter.

1. CONTRIBUTIONS AND AUDIENCE
    (a) What are the important contributions of this paper?

    (b) What group(s) of readers would be interested in the paper?

2. TECHNICAL SOUNDNESS
    (a) Is the paper technically sound?

    (b) Is the mathematical development complete and accurate? If not, please give additional comments.

3. PRIOR PUBLICATION
    Applicable ☐   Not Applicable ☐
    If comparable work has been published, please provide expanded comments to the following questions on a separate sheet:

    (a) Is that work readily available?   Where?
    (b) Does the paper add new insights or interpretations?
    (c) Does the paper serve a tutorial role?
    (d) Are the mathematical derivations useful or are they repetitive?

4. ORGANIZATION AND STYLE
    (a) Is the paper well written and well organized?   If not, how should it be improved?

    (b) Are title and abstract informative and concise?   What improvements should be made?

    (c) Are the methods adequately described so that the work could be reproduced by the reader?

    (d) If the text, tables, or illustrations need to be condensed, please state how.

    (e) Is there adequate reference to previous work?   What references should be added or deleted?

    (f) Is the level and style or presentation consistent with the readership in 1 (b)?

5. EVALUATION
    (a) How do you rate the paper? Outstanding ☐   Very Good ☐   Good ☐   Fair ☐   Poor ☐
    (b) What course do you recommend? Publish ☐   Publish following minor revision ☐   Publish only after major revision ☐   Reject ☐
    (c) Is there anyone else who should see the manuscript before it is accepted for publication?

Figure 5

CARBOHYDRATE RESEARCH

RECOMMENDATIONS TO REFEREES

The editorial staff greatly appreciates the fine work of the referees and the importance of their service. The following information and recommendations may assist in the preparation of reports.

CARBOHYDRATE RESEARCH publishes Full-length Papers, Notes, and Preliminary Communications, dealing with carbohydrates in general, including, for example, sugars and their derivatives (also nucleosides, nucleotides, cyclitols, and model compounds for carbohydrate reactions), oligo- and polysaccharides, glycoproteins, and glycolipids. The following aspects are considered to fall within the scope of the journal: chemical synthesis, the study of structures and stereochemistry, reactions and their mechanisms, isolation of natural products, physicochemical studies, analytical chemistry, biochemistry (biosynthesis, degradation, etc.), action of enzymes, and technological aspects (where these are of interest to chemists in general). Some papers submitted to CARBOHYDRATE RESEARCH may be more suitable for analytical, agricultural, biochemical, polymer, or industrial journals, and in borderline cases the editors will welcome specific recommendations for alternative journals.

Both the editors and the referees should use all reasonable methods to encourage authors to achieve conciseness consistent with clarity in potentially acceptable manuscripts, to reject unsuitable manuscripts, and to recommend submission elsewhere of manuscripts that are more suitable for other journals than for CARBOHYDRATE RESEARCH.

Notes. The standards of quality for notes are the same as for full-length articles. Improved procedures of wide applicability or interest, or accounts of novel observations or of compounds of special interest often constitute useful notes. Notes should not be used to report inconclusive experiments.

Preliminary Communications are reserved for the rapid publication of brief reports of findings of unusual timeliness, and must meet the requirements of significance and urgency.

Recommendations. Please be as specific as possible when recommending revision by the author. With respect to condensation, specific suggestions are desirable as to the amount of condensation needed and its location in the manuscript. Nomenclature should conform to that of *Chemical Abstracts.* Carbohydrate nomenclature should conform with the British-American "Rules of Carbohydrate Nomenclature", *J. Org. Chem.,* 28 (1963) 281.

Sentence structure and good grammar are important. The responsibility for preparation of an acceptable manuscript rests largely with authors. Referees may suggest that papers be rewritten before they are accepted in cases in which good grammar and acceptable style are not followed.

Figure 6

make some points about the ethics of refereeing such as the need for confidentiality, abstention from harsh or abrasive language, and avoidance of personal biases.

Finally, some journals ask for separate, confidential reports to the editors in addition to comments that are intended to be transmitted to authors. Two examples of these reports to the editors are shown in Figures 9 and 10. Both ask for a recommendation on the final disposition of the

CANADIAN JOURNAL OF MICROBIOLOGY

GUIDELINES FOR REVIEWERS

The following guidelines are offered as an aid in reviewing manuscripts for the Canadian Journal of Microbiology.

Please note that the manuscript you have been asked to review is to be considered a PRIVILEGED COMMUNICATION, a confidential document not to be shown, or used, or described to anyone else except to solicit assistance in reaching an editorial decision.

In your comments which are to be passed on to the authors, please avoid sarcasm and phrasing which would antagonize the authors or reveal your identity. If it is necessary to use harsh words, please confine them to the confidential comments. Please make your recommendations as regards the acceptability of the paper only among the confidential comments. Your judgement may disagree with that of other referees and the final decision must be made by the Editor.

1. Is the work scientifically sound?

2. Is the work sufficiently new and significant to deserve publication? Is it significantly different from previous papers by the same author?

3. Are the data clearly and concisely presented?

4. Are the experimental methods, legends to figures and footnotes to tables presented in a manner clear enough to allow other investigators to repeat the work?

5. Is the discussion relevant? Do the conclusions drawn follow from the data presented?

6. Are the references to relevant work adequate?

7. For information retrieval purposes, it is essential that the title of the paper be clear and informative and that the abstract cover all the main findings reported in the paper. Are these conditions adequately met?

8. Is the material presented in the manuscript suited for publication in a microbiological journal, or would it be better suited to another journal, e.g. in the fields of chemistry, botany, biochemistry, pharmacology or physiology?

9. The Editorial Board would greatly appreciate any specific suggestions you might have for shortening manuscripts where this is advisable. Condensation often improves the manuscript but general statements that the authors should reduce the paper by a certain fraction are not very helpful.

10. Comments on matters of form are often very useful. It would also be very helpful if the unnecessary use of nonstandard abbreviations were pointed out. Minor typographical errors or grammatical errors need not be corrected.

Figure 7

ENVIRONMENTAL SCIENCE & TECHNOLOGY

INSTRUCTIONS TO THE REVIEWERS

American Chemical Society
1155 Sixteenth Street, N.W.
Washington, D.C. 20036

Phone: (202) 872-4618

Environmental Science and Technology is devoted to the publication of original research in the fields of water, air, and waste chemistry and with other scientific and technical fields relevant to the understanding and management of the water, air, and land environments.

Contributed research papers will, in general, describe complete and fully interpreted results of original research. Review papers will be considered only when they serve to provide new research approaches or stimulate further worthwhile research in a significant area. In addition to full-length articles the journal will also publish Notes, which are shorter research reports describing preliminary results of unusual significance or studies of small scope.

In attempting to determine the suitability of a manuscript for publication the reviewer should keep the following questions in mind:

- Is the experimental question clearly stated, and is it significant in the context of a known scientific problem?
- Are the experimental data valid and are they consistent with the experimental objectives?
- Are the interpretations of the data sound and are the conclusions validated by the experimental evidence?
- Is the abstract informative and does it give the essence of the research in clear, sufficient terms, i.e., the nature of the problem, the significant data, and the conclusions? The abstract should be self-explanatory and suitable for reproduction by abstracting services without rewriting.

If a manuscript is acceptable, a brief statement describing its significant contributions should be made, together with suggestions for minor improvements, if needed. If the manuscript is recommended for rejection, the major reasons should be stated concisely, preferably in language which can be transmitted to the authors. If a manuscript appears to be acceptable but requires modifications, suggested revisions should be stated briefly. Authors of manuscripts which are declined because of low quality should not be advised to submit their papers to another journal.

The editors do not expect reviewers to rewrite a paper. A manuscript which is poorly written or unclear should be returned to the editor with the simple request that the presentation be clarified before its scientific merits can be reviewed. Papers which, in the reviewer's opinion, are much longer than necessary should be returned with a statement to this effect and, if possible, with specific suggestions for condensation. In this regard, please give particular attention to the possibility of condensing or deleting any of the figures or tables. If you believe that in the interest of saving space some of the experimental details could be deleted from the manuscript and made available upon request in the form of microfilm or microfiche, please so indicate in your report.

Reviewers are requested to refrain from communicating directly with authors or from disclosing their identity without prior editorial consent.

The scientific quality of the Current Research section is strongly related to the quality of the review effort. Because of the diverse subject matter covered by the journal, the editors rely heavily upon reviewer opinion and comments even when the recommendation is strongly for acceptance or rejection. For this reason recommendations that are returned without written comments will be discarded. Manuscript copies should be returned to the editorial office with completed reviews.

If for any reason you should find it impossible to act on a manuscript within the stated time limits, will you, please, suggest other reviewers and return the manuscript promptly to the editorial office.

R. F. Christman, Editor
C. R. O'Melia, Assoc. Editor
J. H. Seinfeld, Assoc. Editor

Figure 8

CANADIAN JOURNAL OF PHYSICS   File No._____

REPORT TO THE EDITOR

In which of the following categories would you rate the paper:

    1. An important contribution to knowledge in the field.
                                                                    _____

    2. A useful contribution though not of fundamental importance.
       _____

    3. A technically adequate paper of limited or highly specialized
       interest.
       _____

    4. A technically adequate paper, but more appropriate for submission
       to another journal. (If possible, please give a specific suggestion.)

    5. A paper too insignificant to justify publication.
       _____

    6. A paper with important technical errors.
       _____

If none of the above categories adequately describes your evaluation of the paper,
please state in your own words how you would classify it.

In the light of the above judgment, do you:

    A. Recommend that the paper be accepted for publication, with at
       most minor corrections or modifications.
       _____

    B. Recommend that the paper be not accepted in its present form (this
       does not preclude the author's submitting a revised version for
       subsequent consideration).
       _____

    C. Recommend that the author(s) be informed that their paper is not
       suitable for publication in the Canadian Journal of Physics.
       _____

                              _____
                                      SIGNATURE

Please return the yellow and green copies.

Figure 9

paper, one (Figure 9) asks the referee to categorize the paper on the basis
of significance, and the other (Figure 10) poses six specific questions on
the evaluation of the paper.

    The selection of one particular system from this rather bewildering
array of guidance and reporting mechanisms is largely a matter of
editorial preference based upon perceived needs. If editors feel well served
by whatever system they are using, then there is no reason to change. On
the other hand, if editors feel that referees overlook an important point

RADIO SCIENCE
INTERNATIONAL JOURNAL
EDITORIAL REVIEW FORM

TITLE AND AUTHOR(S) OF PAPER:

NAME OF REVIEWER:

ASSOCIATE EDITOR TO WHOM REVIEW SHOULD BE RETURNED:

DATE BY WHICH REVIEW SHOULD BE COMPLETED:

Please answer the questions below, and explain qualified or negative responses. Comments for the author should be on a separate sheet, unsigned. If more than one person is involved in the review, they should co-sign below.

Evaluation	Yes	No	Qualified Yes
1. Is the paper technically correct?	( )	( )	( )
2. Does it significantly advance knowledge in its field?	( )	( )	( )
3. To your knowledge is this the first time this work has appeared in print?	( )	( )	
4. Do the references seem to be complete?	( )	( )	( )
5. Are the abstract and title descriptive of the content?	( )	( )	( )
6. Have you made corrections on the manuscript (optional)?	( )	( )	

Recommendation

( ) Publish as is.

( ) Publish with minor revisions shown on attached sheet.

( ) Revise in accord with recommendations attached.

( ) Reject on grounds of _____

( ) Submit instead to _____

Comments on recommendation (not intended for transmittal to authors)

Signature of Reviewer(s) _____ Date _____

_____

Figure 10

consistently or frequently, then it is probably a good idea to remind them of it. For example, if one or two reports fail to mention anything about the adequacy of references, it probably means that the references were satisfactory in those papers and therefore did not require comment. However, if no mention of references is made in reports on say 25 to 30 papers, than it might be assumed that the referees have overlooked that point and a reminder could be in order.

As can be seen from the examples in Figures 2–10, a majority of journals offer at least some guidance to referees, and with the current concerns about the quality of the scientific literature and the ethics of the review process, that is probably a good idea. The directions need not be in the form of a treatise on how to referee a scientific paper. In fact, referees cannot be expected to read more than one page of such stuff; remember, they are anxious to get on with reading the paper that accompanied it. One page should provide ample space to itemize the important points that ought to be considered by referees. At the minimum, those points should cover:

- the traditional constraints on scientific publication
- a few questions about presentation
- one or two gentle reminders about the ethics of refereeing

Something along the following lines would seem to meet the needs and desires of most editors.

GUIDE FOR REFEREES

The following checklist is intended as a guide for referees. It would be helpful if you would comment *specifically*, in your report, on any points in which you consider the paper to be deficient.

Traditional constraints
1. Does this paper report a specific, identifiable, advance in knowledge?
2. Has the work reported in this paper been published before?
3. Are the conclusions justified, soundly based, and logically consistent?
4. Are the procedures and methods sufficiently clear that the work could be repeated by anyone knowledgeable in this field?
5. Are the references to prior work pertinent and complete?

Presentation
1. Is the paper as concise as it could be, consistent with clarity?
2. Referees are not required to comment on writing style. However, they are invited to suggest changes that would remove ambiguity or clarify meaning.

3. Are all figures and tables relevant and properly prepared?
4. Are the title and abstract truly descriptive of the contents?

Ethics

1. We remind referees that the paper under review is a confidential document that should not be discussed or shown to others without the express permission of the editors.
2. As your detailed report may be relayed to the authors as a guide for revising their paper, we request that you avoid harsh or abrasive statements. Comments specifically for the editors should be written in a separate letter accompanying the report.
3. Your anonymity as a referee will be preserved and you are asked not to identify yourself to the authors without the express permission of the editor.
4. Please return your report within the specified time limit.

Of course, a checklist such as that above reflects only the bare bones of refereeing; it says little about responsibilities or style. The referee's role is important and demanding and ideally should be approached with the same care and thoughtfulness required to write a paper. The most satisfactory referees are those who try to be helpful to both editors and authors by giving specific, as opposed to general, comments. For example, a comment that "This paper is too long and could be shortened by one-third" is not very helpful. The editor might suggest that the authors shorten their paper but will probably get back a manuscript that has been retyped, single-spaced with narrower margins, and a letter stating that the paper has been reduced from 28 to 20 pages. Contrast that with a comment, "This paper is longer than it need be. The second paragraph on page 4 and the third paragraph on page 6 could be deleted or reduced to their topic sentences, and the experimental methods on pages 12 and 13 are adequately described in the cited reference." Now the editors and authors have something to go on, something that can be checked in a revision, and the authors have been shown how to be more concise in their writing. Similarly a recommendation for rejection that states simply, "This work is not up to the standards required for publication," puts an editor in the difficult position of appearing to act arbitrarily and does nothing to help the authors raise the standards of their work. If rejection is recommended, the reasons should be specified.

Referees should try to be considerate and polite. The review system for scientific papers works best when all parties — authors, editors, and referees — regard it as an educational rather than an adversarial process. Thus, the goal of referees should not be to shoot down every paper that hits their desks but rather to help the editor show the authors how their paper could be improved. In the case of rejections, the aim should be to explain to authors what they need to do to make their work publishable. These objectives can best be achieved by systematic, logical explanations in moderate language; the use of harsh, abrasive statements serves only to diminish the force of the arguments. The light touch — humor — can be very successful in making a point, as long as it is not sarcastic or cynical. One referee of a physical chemistry paper had glowing praise for the experimental methods and accuracy of the data but was unhappy with the subsequent mathematical treatment. The comment was expressed as follows: "The analysis of such beautiful data by these methods is like trying to play Beethoven's Fifth Symphony on a series of partially filled beer bottles." Few authors could take umbrage at that, but they might if the referee had stated, "The data are fine but the statistical treatment is atrocious."

Referees should be careful not to confuse specificity with nit-picking, particularly in any comments on the authors' use of language. A few people, fortunately not too many, are born nit-pickers and will probably never change. However, it is unproductive for referees to worry about dotting every "i" and crossing every "t"; that should be left for the journal's copy editors. In the matter of language, it is important to remember that the scientific literature is not written in deathless prose that will be studied for centuries for its artistic qualities. The significant lifetime of most scientific papers is no more than 5 to 10 years. Certainly a well-written scientific paper is a thing of beauty, much to be desired, and clear writing is a skill that can be acquired.[6] However, it is not the referees' job to impose their own style upon authors. Thus, comments on the use of language should be confined to those parts of manuscripts that are excessively verbose or where the meaning is confused or ambiguous.

How to Keep Referees Happy

Most scientists who publish with reasonable frequency will, because their names appear in citations and indexes, receive many requests to review papers. Indeed, the number of requests can become a flood, and it is by no means unusual for an individual scientist to receive, at certain times, simultaneous requests to review up to six papers from as many different journals. At such times, most scientists will return some of those six papers with a note that they are too busy to review them; they may or may not explain that the paper being returned was the fifth or sixth request for review that they had received in two days. Clearly, referees in those circumstances will exercise some sort of selection process. They will probably give top priority to papers from those journals in which they themselves publish most frequently. However, beyond that the referees are most likely to be influenced by their past experiences with different editors. It is therefore worthwhile for editors to consider how best to maintain good relations with referees. The key elements in keeping referees happy are

- Do not overload them with excessive demands.
- Provide feedback in the form of information about the consequences of their reports.
- Show appreciation.

Do Not Overload Them

What constitutes an overload or excessive demands? Probably everyone would agree that requests to review a paper a week or even every two weeks would be excessive. Even one paper a month (12 per year) might be regarded as a heavy load. Editors must remember that referees are not theirs exclusively and receive papers for review from several journals. That could easily translate into three or four papers per month if each journal allowed only a one-month interval between requests. It would therefore seem that an interval of two or three months (four to six papers per year) is reasonable and not likely to raise the hackles of many referees. Obviously, one editor has no control over what other editors are doing, and some referees (particularly the good ones) will still end up with three or four simultaneous requests for reviews from different journals. However, if those referees know that the editor of journal A has a policy of maintaining an interval of two or three months between requests for review, they are likely to be more favorably disposed towards that journal than towards journal B whose editor does not seem to recognize any time interval. Thus, it is a good idea for editors to decide on a minimum interval to be maintained between requests for reviews from the same referee and to declare it, either in an editorial published in the journal or in the standard letter that requests review of a manuscript. Of course, following Murphy's Law that, "if anything can go wrong, it will," an editor will no sooner declare a minimum interval between requests for reviews than an exception will arise. It may be that an author has spent a sabbatical writing up three years' work and submits four or five substantial and interdependent papers. Or, a paper may be submitted that is a perfect match with the interests of a referee who just sent in a report on another paper two weeks ago. When that sort of thing happens, exceptions to the minimum interval can be made once, but only once, with a special letter of explanation, or a telephone call, and a guarantee (which must be honored) of subsequent relief from refereeing for a certain period. Possible letters to cover such situations might read as follows.

Dear Dr. Angler:

The enclosed manuscript, "The Wiggling Coefficient of Worms on Fish Hooks," by Fisher and Hooker, has been submitted for possible publication in our journal. We note from our records that you reviewed another paper for us recently. Normally, we try to allow an interval of three months between requests for a review. However, the subject matter of this paper seems so close to your interests that we thought you might enjoy reading it and we were certain that you would be the best qualified referee. If you would be prepared to review this paper we will double the interval between requests and will not send you another manuscript for six months.

Thank you very much for your assistance.

Editor

Dear Dr. Short:

I am writing to seek a special favor from you. Drs. Long, Stem, and Stern have submitted the five manuscripts enclosed, "Functions of the Vertebrae, Parts 7–11," for possible publication in our journal. Because these five papers are closely related and interdependent it seems essential that they be reviewed together; matters of overlap, repetition and undue fragmentation are of obvious concern. We would be most grateful if you could take on this unusual refereeing chore for us. In return, we promise to refrain from sending any further requests for reviews for one year.

Thank you very much for your assistance.

Editor

Provide Feedback

Referees appreciate feedback, particularly when they have worked hard on a difficult or complex paper. It is therefore a good policy, when referees have provided substantial, constructive criticism, to let them know about the consequences of their reports. Whether or not a referee should be asked to check a revised manuscript is a matter of editorial judgment for which no specific guidelines can be given. When editors feel that authors have met substantial criticisms by major revision, a note to the more critical referee along the following lines might be appreciated.

Dear Dr. Judge:

About six months ago you reviewed a paper for us, "The Rate of Formation of Dew Droplets as a Function of Temperature," by U. R. Damp and I. M. Cold. You had some major criticisms that we required the authors to meet before considering their paper further for publication. They have now sent us a manuscript that has been extensively revised according to your comments. Because your report was clear and specific, it was easy to check the revision and I am satisfied that your major objections have been met satisfactorily. There seems no need to bother you further with this paper. I just thought you might like to know what had happened and it gives me a chance to thank you again for your assistance which has resulted in a greatly improved paper.

Yours sincerely,

Editor

A sure way to give referees dyspepsia is to have a paper appear in print after they have recommended rejection. There will be a certain number of papers, particularly in controversial areas, about which referees will disagree. Although one referee may recommend rejection, the editor may find the authors' rebuttal and the other referee's report more convincing. When that sort of situation arises, the more critical referee deserves an explanation, perhaps along the following lines.

Dear Dr. Slasher:

Recently you were kind enough to review a paper for us, "The Warp and Woof of Continental Drift," by Plate and Shelf. You had some major objections to this paper and recommended that we turn it down. As sometimes happens, the other referee was less severe, so we gave the authors the chance to respond to your report. Copies of that response and of the other referee's report are enclosed. I have decided that the paper warrants publication after revision along the lines proposed by the authors. It seems that their new observations are soundly based and may help to resolve the legitimate controversy in this area.

I just thought that you might like to know what had happened with this paper and that your advice was not ignored without careful consideration.

Thank you again for your assistance,

Yours sincerely,

Editor

The reaction of referees to this kind of explanation is best illustrated by the following three excerpts of letters from the files of an editor who believes in feedback.

Thank you very much for your letter regarding the manuscript that I reviewed recently. It was kind of you to take the time to write and explain the situation to me.

I'm not upset by your decision in any way. In fact, the author's rebuttal makes quite clear the synthetic utility of the reaction. This fact escaped me in the original manuscript. Again, thank you for writing.

Thank you for your courtesy in letting me know the final outcome on the paper that I reviewed recently. This is just to let you know that

I have no difficulty in accepting your final decision. It is all part of the game; I played my role as best I could, but in your position I might well have decided as you did.

> Thank you for your letter outlining your handling of the manuscript I refereed.
> These types of decisions ultimately and rightfully rest with the editors, of course, as they are responsible for the philosophy of the journal. I do appreciate your taking the time and effort to let me know the results of my report, and the thinking that was behind it. This was the first instance that an editor has been sufficiently thoughtful to do this.

Another good way to provide feedback to referees is to send them photocopies of the other referee's report and of the editor's decision letter to the authors. This procedure has several advantages: it lets referees know that their reports have been received, shows them how the editor has reacted, and permits them to compare their report with that of the other referee. Referees can therefore see if they have been too lax or missed some crucial points. They also develop an understanding of the standards required by the journal and the need to meet those standards in their own work. The routine exchange of editors' letters and referees' reports costs a bit in postage and secretarial time. However, it requires no effort on the part of the editor and provides an ideal mechanism for all the feedback that is necessary.

Show Appreciation

Referees do not expect a personal letter from the editor in response to every review. However, they do like to know that their report has been received and that their work is appreciated. Thus, every referees' report, without exception, should be acknowledged in some way. There should at least be a postcard or a form letter that states, "This is just to let you know that we have received your report on the paper '_____,' by _____ and _____. Thank you very much for your assistance."

Some journals show their appreciation of referees by publishing a list of their names in the journal, once a year, with a note of thanks. Although the publication of referees' names would seem to violate the principle of anonymity, they are only listed as having refereed for the journal; no attribution by paper is or should be given. It would be difficult for anyone to associate a particular paper with names in a collective list of two or three hundred referees. In addition to being a pleasant courtesy, the publication of a collective list of referees' names also provides the readers of the journal with the opportunity to get some idea of the scope and quality of the refereeing and in that sense is good public relations.

Referees provide extraordinary, voluntary services that are vital to the very existence of scientific journals. Most scientists are fully prepared to act as referees without thought of recompense; they correctly regard refereeing as part of their professional responsibilities. Nevertheless, those same scientists will be more willing to review for journals that show some appreciation of the referees' contributions. It is therefore in their own best interests for editors to maintain good relations with referees: keep to a stated, minimum interval between requests for reviews from the same referee; provide feedback when a referee's advice is not followed or when reports have been particularly helpful; acknowledge every report and say "thank you."

Chapter 6

Ethics

Ethics are the principles of morality, the rules of conduct, or the duties and obligations associated with a particular activity or association of people. Unlike laws, ethics are seldom spelled out in any detail; indeed, they may change over time as variations occur in the conditions and attitudes that affect human relations. Ethics are essential to all human activities in bridging the gap between punishable crime and chaotic laissez-faire. Ethics are taught largely by example although the most generally applicable ethical principle, the Golden Rule (do unto others as you would have them do unto you), is part of the formal teachings of nearly all religions. Ethics are enforced, not in any formal sense, but by the social pressure that the affected community applies to unethical behavior.

Scientific publication has had its code of ethics from the beginning, starting with the normal constraints of significance, testability, repeatability, and due reference to prior work. For the first two centuries of modern science those clear-cut ethical principles, enforced by editors and referees, were fully sufficient and seem to have presented few problems. In more recent times, however, some other issues have arisen as a consequence of the proliferation of science, greater involvements of governments, exploitation of science through technology, and development of complex funding mechanisms. Price[23] has suggested that the intuitive ethics of the past may no longer be valid or sufficient and that a rational ethic is needed, based upon new knowledge about the use of scientific information. Many of the new questions of ethics arise from the pressures to publish or perish and involve such issues as citations and their use, multiple authorship, and duplicate publication. While these kinds of issues should be addressed by the scientific community as a whole, through societies, federations, or international unions, it is obvious that primary journals have an important role to play. The commitment of this chapter to ethics is not intended to imply that unethical behavior is a widespread or frequent occurrence in scientific publishing. However, current concerns about ethics in science, as expressed in the popular press and letters to editors of some journals, suggest that some attention is needed. The following discussion simply outlines some ethical problems that may arise in scientific publication and what can be done about them. As in other areas of human activity, the ultimate responsibility for ethical behavior rests with the individual, and it is sometimes difficult to see what other kinds of pressures or controls can be applied.

Authors

Fraudulent Results

The authors of scientific papers clearly bear the full responsibility for the veracity of the work reported therein. As a general rule, referees and editors do not question experimental results; indeed, it might be regarded as unethical for them to do so. If a question is raised, it is likely to be a very tentative suggestion such as, "The result reported on p. 7 seems somewhat improbable. Are the authors sure that it is not an error?" Does this mean that authors are at liberty to publish fraudulent results with no checks or accountability? Not at all. The practice of science is self-policing in that respect, and the consequences can be severe for anyone who intentionally publishes fraudulent results. Such persons will be ostracized by the scientific community and, unable to obtain positions or funding, their scientific careers are finished. Will frauds always be caught? Almost certainly, particularly if they involve a sufficiently important topic. Results on an important, significant topic will attract attention as soon as they are published, and attempts to repeat and use the work will be made in several laboratories. Once it is found that the results cannot be reproduced, the word goes out and the offending authors will be at least discredited in the eyes of their colleagues. The authors may be questioned discreetly about possible inadvertent omissions in their experimental protocols, and if no satisfactory explanation of the irreproducibility can be found, the falsifier of information will eventually be tracked down. What happens when the falsified or fraudulent results are not significant? Nobody pays any attention, of course, and the paper just sits there, unheralded, untested, and uncited. In short, if the fraud is significant it will be detected quickly; if not significant it will be ignored. It is difficult to believe that the literature is loaded with "fudged" results that arise from the pressures to publish or perish. The ease of detection and the high risks involved are just too great. Furthermore, the many thousands of papers that have been used successfully and cited in further work are testimony to the reliability of the record. Science could not have progressed as it has if the published record were significantly unreliable.

Although the identification and banishment of the perpetrator of a fraud is the major consequence of such unfortunate events, there are peripheral effects that involve the responsibilities of authorship. Thus, principal investigators who have coauthored papers that, unknown to them, contain falsified results, have an obligation to withdraw the papers, publish a retraction, and try to assure their colleagues that steps have been taken to prevent any recurrence. Editors should certainly cooperate with authors who find themselves in such an unfortunate situation. The prevention of fraud is really a matter of organization and supervision in the laboratory. The story is told that, in Emil Fischer's laboratory, the first task given to a new graduate student was to repeat the experimental work in the completed thesis of a recent graduate. The benefits of this procedure were manifold: (1) the new student became familiar with the techniques of the laboratory, (2) a further supply of new compounds was obtained, (3) the work was verified as being reproducible, and (4) the new students became aware that their original work would subsequently be checked. We are also told that Fischer did not publish until the verification procedure was completed.

Multiple Authorship

Multiple authorship poses another ethical problem that is primarily the responsibility of authors. The trend towards team research continues to increase, and multiauthored papers are now more the rule than the exception. The ethical question is, of course, who warrants coauthorship? Authorship is a principal reward of science and is generally regarded as a measure of success and achievement. Garfield[11] has discussed the issue of author attribution at some length and has suggested that journal editors, collectively, should formulate some guidelines for authorship and ask authors to affirm that they qualify. A comparison was made to patents for which inventors must sign an affidavit declaring that they are the originators of the invention. Since authorship implies responsibility, one simple guideline could be that all authors should be capable of participation in a discussion or defense of their paper. This requirement would eliminate most support people such as technicians, data gatherers, computer programmers, and administrators who presumably would not be able to deal with the concepts and scientific implications of the work. It might also eliminate the senior scientist, in charge of a large group, who has not maintained close enough contact with the laboratory to be familiar with the latest experimental methodology. For multidisciplinary papers, the specialty of each author could be defined in a footnote so it would be clear which authors could discuss the different aspects of the paper. For example, a paper that reports the discovery of a new antibiotic might well involve microbiology, chemistry, and pharmacology. It would not be reasonable to expect, say, the chemist author to be able to discuss the detailed aspects of the other two specialties, but each author, since they worked and published as a team, should be knowledgeable in the broad concepts of the work.

A guideline for authorship, if presented by journal editors collectively, might exert some moral pressure but would be difficult to enforce. Who would check to see whether all authors met the guidelines? What penalty would or could be imposed for violations? The situation is not the same as with patents where the inventors reveal their understanding of the invention during preparation of the patent application and where there is legal recourse, under the patent law, for violations. It would be sad to think that the formulation of guidelines might lead to a proliferation of litigation about authorship rights. Authorship attribution is a truly ethical issue and should be left that way; there are risks to over-regulation. Thus, guidelines might be suggested as a form of assistance but not made obligatory because, in the final analysis, the decisions about authorship properly belong to the individuals involved.

A fine example of the ethics of authorship was experienced by one editor a few years ago. One referee of a paper in synthetic organic chemistry disagreed with a mechanism that the authors proposed to account for the formation of compound C from compound A under certain conditions. The referee suggested another mechanism as being more probable and noted that this latter mechanism could be proved if compound B, which could be prepared rather easily, also gave compound C under the same conditions as from compound A. The authors wrote to the editor that the referee was very probably correct and that they would do the experiments required. A few months later a revised manuscript arrived with the description and proof of the mechanism suggested by the referee. In a covering letter the authors stated that they would like to invite the referee to be a coauthor for having suggested that mechanism and the way to prove it. The editor relayed that letter, with the revised manuscript, to the referee, who responded that the authors had done all the work, that the project was theirs originally, and that coauthorship was not warranted. The authors then asked if the referee would agree to let his contribution be recognized in an acknowledgment. This the referee accepted, modestly and somewhat reluctantly, when the editor pointed out that his suggested mechanism was an important part of the paper and should be credited to him. Such examples of high ethical standards bring joy to the heart of an editor and make it all worthwhile.

The order in which authors' names are listed on multiauthored papers is a more difficult issue because practices vary from lab to lab, from discipline to discipline, and from journal to journal. There probably will never be agreement on any system for ordering authors' names and, certainly, it is impossible to draw up any ethical guidelines based upon current experience. Fortunately, the variations in ordering names present difficulties only to those administrators and citation analysts who try to assign "credits" to individuals on multiauthored papers. With very rare exceptions, scientists' reputations are not based on a single paper. Scientific stature and recognition come from being an author on a succession of high-quality papers. In such a succession the names that recur, regardless of order, are the ones that will be remembered and become known.

The increasing frequency of multiauthored papers has given rise to another problem: the reluctant author. At some time most editors have received letters from authors along the following lines: "Dear Editor: I have learned recently that a paper has been submitted to you with my name listed as a coauthor. I should like you to know that I had nothing to do with the preparation of that paper, disagree with the conclusions, and insist that my name be removed." Editors should, of course, accede to such requests and, in dealing with the author who submitted the manuscript, should note that it is unethical to list as a coauthor someone who does not want to be associated with the paper and who may not even have seen the manuscript. Although this issue is clearly the responsibility of authors, it is relatively easy to control if any journals find it to be a significant problem. When a manuscript is submitted, acknowledgment of its receipt can be sent to all authors, not just the corresponding author, particularly when authors have different addresses. A further check exists where copyright law requires that authors sign a release when an article is accepted for publication.

Plagiarism

Another ethical issue that primarily concerns authors is plagiarism, both intentional and unintentional. Intentional plagiarism, the verbatim or near verbatim transcription of an already published or submitted work without attribution, is a highly unethical act that has the same harsh consequences as publication of falsified results. When intentional plagiarism is detected, editors have some particular responsibilities that are discussed later in the section on ethics for editors.

Unintentional plagiarism occurs when authors use, without attribution, ideas, concepts, or results that have been published previously or were communicated personally at meetings, seminars, or conferences. Garfield[11] has called this kind of plagiarism "citation amnesia," and it arises through oversight, forgetfulness, or lack of thoroughness by authors. Although referees are asked and expected to comment on the adequacy of references in a paper, it is the ethical responsibility of the authors to make sure that all pertinent references to prior work on which their paper depends are included. A literature search should routinely be the initial step in preparing a paper. With the current availability of abstract and index services and computerized retrieval of information, the omission of this important step is inexcusable.

Duplicate Submissions

It is not uncommon nowadays for an editor to shepherd a paper through review and revision, pass it for printing, and then, or even at the galley stage, receive a request from the authors that the paper be withdrawn. Sometimes there are good reasons for withdrawal; the authors may have detected an error through subsequent work or may want to revise their ideas in the light of a recently published paper from another laboratory. However, sometimes there is no explanation given for the withdrawal, and the editor is then surprised to see the withdrawn paper appear in another journal within a few months. What has happened is obvious. With the pressures to publish or perish (again) the authors have submitted their paper simultaneously to two or more journals thinking that one, at least, may accept it and that time can be saved by avoiding the submission, rejection, resubmission route. If, perchance, more than one journal accepts the paper, then it can be withdrawn from those that are either slower to publish or less prestigious in the authors' eyes. Duplicate submissions must be regarded as unethical because they impose a needless expense and a waste of time on the editors and referees of the journals involved and violate a policy that is clearly stated in most journals. There is nothing wrong with the submission of a rejected paper to another journal because, as already discussed, journals vary in their definitions of acceptability. However, simultaneous submission of the same paper to two or more journals is a deception. Some journals require authors to sign a statement that a submitted paper has not been offered to any other journal. Such a statement could presumably be used to recover expenses from violators, something that editors should be urged to do when they have evidence of simultaneous submissions. Some journals have "blacklisted" authors who were found to have made duplicate submissions; the editors refused to consider further papers from those authors for a period of three years.[3]

Fragmented Publication

Fragmented publication represents a less serious ethical problem for authors because there are often justifiable reasons for publication of the same work several times in somewhat different forms. Consider the scientists who have made a really important, significant discovery in an active, competitive field. They will want, quite justifiably, to publish a brief, quick announcement of their discovery to establish priority and to let others in the field know about their results. That announcement will be in the form of a Communication or Letter to the Editor. The authors are then invited to present their work at a conference and, of course, are required to produce a manuscript for the Conference Proceedings. However, the authors still have one or two experiments left to finish off the work properly, and besides, they do not want their definitive paper to be buried in an obscure volume of a conference proceedings. They therefore prepare a thorough, definitive manuscript for publication in a reputable primary journal. One of the authors may then be invited to be a principal speaker at an international conference, and the work will be published yet again in a volume of plenary lectures. Finally, the authors will be asked by some publisher to write a review, and since their work predominates the field, the review will consist largely of a reworking of their major paper and those that led up to it. There is no denying it; that is how the system works, in many cases quite justifiably. The problem is that the system is open to abuse, and there are no obvious ways to control it. The presumption that each and every one of the thousands of Communications or Letters that are published annually contain important, significant discoveries stretches the bounds of credulity. A significant portion of those kinds of preliminary announcements and subsequent short papers, the L.P.U.'s (least publishable units[2]), are published to build up authors' bibliographies for career advancement. One editor, who had been trying for some time to reduce an author's undue fragmentation of results, finally received a letter from the miscreant along the following lines: "Incidentally, you will not have to worry any more about those short papers from me. I was just granted tenure and can now afford fewer papers." Unethical? Perhaps, but is the author really to blame for doing what seems to be demanded by the system and is done by everyone else? As stated earlier, but worth repeating, the pressures to pad bibliographies will only be eased when grants panels and tenure committees give clear demonstrations, by their actions, that a few substantial, definitive papers are more highly regarded than a long list of Communications, Letters, and L.P.U.'s. In the meantime, journal editors can do little more than they are doing already: insist upon high standards of significance for Preliminary Communications, press authors to publish completed studies rather than a succession of L.P.U.'s, and take a very hard stand against publication of anything that has appeared before in conference proceedings or the like.

Referees

Basically, the ethical responsibilities of referees are covered by application of the Golden Rule: referee papers as you would wish others to referee yours. The substance of refereeing was discussed in Chapter 4; the courtesies and behavioral aspects were given only passing mention.

Delay of Reports

All authors want and expect swift action from referees, but when the positions are reversed these same impatient authors, now referees, will let a manuscript sit on their desks, ignored, until some long-suffering editor is forced into additional postal or telephone charges in an attempt to extract a review. Referees should make every effort to review manuscripts within the period specified by the journal, simply as a matter of courtesy and because they expect referees of their own papers to reciprocate. Some delays are inevitable. Scientists move around a lot, and papers for review do not fit well into a travel itinerary. Good referees in a popular field may receive four or five papers for review from different journals simultaneously. There are many valid reasons for delay on the part of referees. Editors do not expect the impossible but they do appreciate knowing what is going on. Referees who are traveling should have somebody watch their mail with instructions to return promptly, with an explanatory note, any manuscripts that were sent for review. Most journals ask that reviews be returned within two weeks of receipt of the manuscript. If referees know that it will take them more than one or two weeks longer than the deadline to complete their reviews, then they should return the manuscripts immediately. However, a one- or two-week delay is acceptable because at least that much time would be taken in the mails to return the manuscript and send it out to another referee. In cases of a delay of one or two weeks, it is helpful if referees telephone or write to the editor to explain what has happened. The editor is then able to write to the author:

> Dear Dr. Ready:
>
> This is just to let you know that there will be a modest delay in getting a second report on your paper, "Rapid Rates of Reproduction." One of the referees has written that his report will be delayed about 2 weeks by a case of whole-body poison ivy, contracted after a midnight skinny dip. He has promised to do his review as soon as he stops scratching.
>
> Yours sincerely,
>
> Editor

Such a letter lets the author know that the editor is paying attention and also provides some assurance that the referee is not sitting on the manuscript for some nefarious purpose.

It is, of course, highly unethical for a referee purposely to delay a paper. Authors touched with paranoia will claim that their papers have been delayed by referees who want to scoop the results or filch the ideas. Editors will state quite bluntly that referees rarely delay a manuscript on purpose; it simply is not a practical thing to do. Except in very unusual coincidences of timing, it takes too long for referees to filch either results or ideas, incorporate them in their own manuscripts, and get them published before the authors' papers. To maintain that time differential, editors must have a firm period beyond which they will not wait before taking action on a paper. If referees have not responded within that time despite all proddings, then editors should proceed immediately with the manuscript, either by relying on the one report that they presumably have or by arranging for a quick review by an associate editor or a colleague.

On rare occasions, referees may receive a manuscript for review when they themselves have a paper already prepared on a closely similar or identical topic. When that happens referees have three options, all ethical:

1. They can return the manuscript to the editor immediately and claim disqualification as a referee because of a conflict of interest; they can then proceed with independent submission of their own paper to another journal.
2. They can return the manuscript, not reviewed, together with a submission of their own paper and ask the editor to consider simultaneous publication if both papers prove to be acceptable.
3. Through the editor, they can exchange manuscripts with the authors and suggest joint publication of one paper.

Whichever option is followed, speed is of the essence; any delay will see the referees suspected unjustly of plagiarism.

Confidentiality

Ethically, referees are expected to treat manuscripts, sent to them for review, as confidential documents. After all, mistakes or errors may be found and the presentation may be altered considerably before publication. Until a paper is accepted, it is the private property of the authors and should be treated as such. Referees should remember that they are confidential advisors to the editor and, just as editors maintain the confidentiality of referee's reports, so should referees respect the confidential nature of the authors' manuscripts. That means that referees should not circulate or photocopy any papers that are sent to them for review. Nor should those papers be discussed with others in the referee's group. If a referee thinks it desirable or necessary to discuss a paper with a colleague, then the editor should be informed and the colleague should cosign the referee's report. All copies of manuscripts should be returned to the editor with the reports or destroyed in those instances when journals do not require return of manuscripts. The ethic of confidentiality on the part of referees is straightforward and is perhaps the clearest application of the Golden Rule: if you expect confidentiality as an author, then you should practice it as a referee.

Incorporation of Ideas

Do referees pick up ideas from manuscripts that they have reviewed and incorporate them in their own work? Yes, probably. The ethical question is whether the ideas are incorporated from a manuscript intentionally and immediately by the referee, or whether they are recalled from a subconscious storehouse of ideas at some later stage. Of course, it is unethical to incorporate ideas from another's manuscript directly without attribution. At the very least, a waiting period should be observed until the reviewed manuscript, from which the ideas came, is published and can be cited. Ethically, referees who want to use ideas from a manuscript they have reviewed should explain their interest to the editor and ask how this interest might be communicated to the authors. Referees might want to suggest a collaboration or ask permission to refer to unpublished results with appropriate acknowledgments. Editors should strive to maintain referees' anonymity in such cases, but it may not always be possible.

There is no way for even the most conscientious individuals to avoid the unintentional use of ideas contained in manuscripts that they have reviewed. Scientists are bombarded by ideas and concepts from a variety of sources: group bull sessions, discussions at meetings, conferences, and seminars, readings in journals, grant proposals, and papers for review. If an idea is not used soon after it is learned, its origin will be difficult to recall and impossible to document. The best that the ethical scientist can do is to search the literature for a citable reference; if the idea came from a paper that was reviewed more than a year previously, the chances are that it will have been published.

Editors

Confidentiality

Possibly the most important ethical responsibility of editors is to preserve confidentiality in their review systems. Authors' manuscripts are their private property and should not be shown to anyone other than confidential referees. Similarly, referees' reports and editors' correspondence are private papers that should not be shown or made available to anyone else, even inadvertently. Thus, manuscripts and all documents pertaining to them should be kept in covered files and should not be left around open or unattended in an open office. When not in active use, all files should be kept in secure storage. Such precautions may seem excessive, but consider the impression on visitors to the editor's office. If they should see open files and copies of correspondence lying about, accessible to any casual passerby, they will not have much confidence that their own papers will be treated any differently.

Editors should deal exclusively with authors and should not get involved with any bureaucracy that happens to be imposed by the authors' employers. Some institutions maintain internal editorial offices to pre-review papers and to assist authors in preparing their manuscripts. Sometimes the heads of internal editorial offices tend to adopt a control function, submit all papers from that institution, and ask that referees' reports be returned to them. Editors can accept the submissions because the authors can be informed by sending each of them an acknowledgment of receipt. However, on ethical grounds, editors should steadfastly refuse to return referees' reports to anyone other than the authors. If some institutions require scientists to reveal referees' reports, that is their business as part of their employer–employee relations, but it has no part in the editorial process.

Unethical Experiments

The issue of ethical experimentation arises primarily in the biological and biomedical fields. It involves the proper care and treatment of animals and the use of humans in medical studies, clinical trials, and drug tests. Although the ethics of experimentation are primarily authors' responsibilities, journals and editors have a role to play. First of all, no publisher or editor would like his or her journal to become known as a repository for papers which report unethical experiments. Secondly, by requiring authors to conform to guidelines, journals and their editors can actively assist those organizations which are concerned with the development and enforcement of the ethics of experimentation with animals or humans. Editors of journals that are affected by this issue should make sure that references to appropriate Guidelines or Recommendations are given in their Instructions to Authors. Specific instructions should be given that experiments reported in the journal are required to conform to those Guidelines, and referees should be invited to pay particular attention to that issue. Editors of biological and biomedical journals might well consider serving a term on one of the councils or commissions that formulate the Guidelines for Ethical Experimentation. At the very least, there should be some liaison to keep editors informed of any changes.

Fraud and Plagiarism

The publication of falsified results or direct copies of previously published papers is an event that all editors hope to avoid; nobody wants to be editor of Acta Retracta or Acta Errata. If, despite eternal vigilance, a fraud or plagiarism should slip through the review process, editors have an ethical responsibility to see that the record is set straight and to inform the innocent who might be adversely affected.

In the case of falsified results that have been revealed after publication, editors should insist that a retraction, correction, or withdrawal be published, either by the authors or as an editorial announcement.

There are several precedents for editorial action following the victimization of a journal and its readers by an intentional plagiary. The editorial board of the *Journal of Infectious Diseases* apologized to their readers when it was found that at least five of the six figures in an article "were taken from the previously published work of other authors."[7] That precedent was cited by the editorial board of *Science* in a similar apology for what they called "An Unfortunate Event."[8] The *Canadian Journal of Physics* published a paper that was substantially the same as one that had been published two years previously by different authors in *Progress in Theoretical Physics*; an explanatory editorial note was published with the statement, "We regret that the similarity was not noticed before publication."[9] More recently, in an article that revealed how several published papers had been pirated, Broad[1] recommended that, "for journals that have printed papers now known to have been pirated, retraction seems like the logical move." At about the same time, a letter to the editor of *Nature*[24] announced the retraction, from the *Japanese Journal of Medical Science and Biology*, of a paper by the same author discussed in Broad's article.

The message in these examples seems clear enough. Despite the best efforts of editors and referees, even the most prestigious journals can become victims of "unfortunate" situations. When that happens, the publication of an explanation, with an apology or retraction, protects the integrity of the journal, clarifies the record, does justice to the innocent, and may act as a deterrent against similar incidents in the future.

A case of potential plagiarism that was stopped by a vigilant editor may be cited here to reemphasize how important it is for editors to look up and read papers when searching for non-obvious referees. The editor in question received a paper on a somewhat obscure topic in physical chemistry. No obvious referees were turned up by the references or by current subject indexes. The editor therefore sent the paper to two generalists, both of whom were known to be acute and perceptive physical chemists. These referees had some minor criticisms but also expressed their unfamiliarity with the topic and noted that there seemed to be little background literature. The editor, not satisfied with those reports, went to the subject index of *Chemical Abstracts*, to search out a more knowledgeable referee, and there found a pertinent reference some 10 years old. The editor looked up that reference and found a remarkable resemblance to the submitted paper under consideration. In fact, detailed comparison showed that the submitted paper was a verbatim transcript of the paper published 10 years previously, by other authors, in another journal. Of course, the editor rejected the paper, with a stern letter to the author about his unethical behavior, and also informed the head of the department (where the author was a visiting scientist), the dean of the author's home institution, and the editors of all journals in which the author had published previously. Note to editors: to benefit from the literature, you must read it.

Chapter 7

Keeping Track

The Need for Records

Authors feel parental about their manuscripts, and with good reason. Manuscripts are the scientist's lifeblood; they contain the hard-won results of daily research over an extended period and, when published, represent the contributions to knowledge by which their authors are judged. Submission of a manuscript may be likened to sending a child to a camp for the first time. The parents want to know that the child is being cared for, is not lost, and can be contacted in emergencies, and that the camp is well run. Experienced editors understand these feelings and recognize that authors are concerned only about their own manuscripts; to the individual author, all those other 40 or so manuscripts that are in the system might just as well not exist. Thus, editors should put themselves in a position to show authors that manuscripts are being cared for and are not lost, that a status report is instantly available, and that the editorial office is efficient. To achieve these ends, everything having to do with a manuscript needs to be acknowledged and recorded. An acknowledgment of everything received in the editorial office lets the sender know that it was not lost in the mails and that someone is paying attention to its arrival. Complete and updated records provide instant information about the status of any manuscript.

The Editorial Assistant

Every journal's editorial office requires an editorial assistant, even if only on a part-time basis. There are undoubtedly some small journals whose publishers feel that they cannot afford such assistants. Such an attitude is false economy that threatens the very existence of the journal. The journal that is edited from an office in a family home is rapidly becoming a relic of the past. Anyone who tries to edit a journal under those circumstances certainly deserves admiration and respect. However, they simply cannot compete with professionally run journals in terms of efficiency. Furthermore, editors who have to deal with all office routine and record keeping, in addition to finding referees, making decisions, and communicating with authors, will certainly not have time left for their own research. Their journals can then expect to reap the consequences of having an out-of-touch editor: loss of confidence and credibility in the eyes of authors and referees. A journal in such circumstances is on a downward trend that is not easily reversed because it will be difficult, if not impossible, to persuade any good, active scientist to become its editor.

While the scientific quality of a journal depends upon the editor, the efficiency with which the journal is run depends almost entirely upon the editorial assistant. The selection and appointment of this person should therefore receive high priority; it is among the most important decisions that an editor will make.

The desirable qualities for an editorial assistant are good organizational ability, good communication skills, typing ability, discretion, and a pleasant personality. The editorial assistant is, in effect, the office manager for the journal, but with all of the office functions embodied in the one person.

Organizational Ability

Editorial assistants should be able to organize the editorial office to facilitate the smooth and efficient flow of manuscripts through the review process. This involves the setup and management of manuscript files, referee files, statistical records, the follow-up system, and all those items involved in manuscript tracking. Most of the action in the editorial office is initiated by the editorial assistant, be it opening and recording new submissions, reminding referees, or making sure that mail gets answered. Editorial assistants should therefore be vigorous self-starters who can keep things moving, including their editors. A realistic schedule should be developed to keep the paper flowing through the editorial office; good editorial assistants should not hesitate to prompt editors who have forgotten to look at their editorial "in" basket for a day or two.

Communication Skills

Editorial assistants need good communication skills for dealing with inquiries that come in by telephone or letter. Authors will call or write to ask about the status of their manuscripts, new authors may have questions about journal format, referees may request clarification, and copy editors or printers may need guidance and advice. There will also be outgoing messages such as reminders to tardy referees and inquiries to authors about revised manuscripts that are overdue. Most of the correspondence will be addressed to the editor, who should see every letter without exception. However, many editors will delegate to their assistants the responsibility of responding to routine inquiries that do not involve science.

Typing Skills

Editorial assistants will be expected to transcribe the editor's correspondence whether it is handwritten, dictated, or recorded, and that, together with the need to maintain records, makes the ability to type a mandatory requirement. Depending upon the editorial office, the extension of typing skills to the operation of word processors or small computers may be a desirable attribute.

Personality

A journal can be counted among the blessed if the editor's assistant has a pleasing personality. The person in this position is, in fact, the receptionist for the journal, the first point of contact, and the impression given to a visitor or a caller will be a lasting one. If the editor's assistant is courteous, pleasant, helpful, and informed, the journal will acquire a reputation as being pleasant to deal with and efficient. The importance of such a reputation should not be underestimated; it can be a major factor in attracting and keeping good authors and willing referees. With the multitude of journals being published today, authors and referees can afford to be selective and will quickly desert any journal if they encounter rudeness or inefficiency.

Recruitment

How does an editor go about finding a suitable assistant? The answer is, of course, use money! People with the qualities desirable in an editorial assistant command good salaries and deserve them. This is not the place to economize; somebody has to do that work, and if it is not done well, the journal will be in trouble. Candidates can be recruited either by direct advertising or through the employment branch of the editor's home institution. The salary offered should be at whatever the local rate is for a good administrative assistant. The editorial offices of small journals can probably get by with part-time assistants. In most places, it is relatively easy to find suitable people who are interested in working half days or others who might take on a bit of overtime work, paid an hourly rate or a negotiated sum per manuscript handled. Whichever method of payment is used for this part-time work, the amount should be sufficient to attract high-quality candidates.

Form Letters and Cards

Much of the routine exchange of information in editing a journal can be handled properly and conveniently by the use of form letters or postcards. In fact, for the majority of papers, the only letters that need to be drafted individually are the editor's decision letters to authors after referees' reports have been received.

Form letters can be either printed in advance or entered into a word processor from which printouts can be obtained on demand. Postcards are convenient to acknowledge receipt of items that do not require a letter response and can be color coded to reduce the possibility of error. They should all be designed so that the specific information is easy to add: names and addresses, titles, dates. The following examples demonstrate some possible form letters and the occasions when they can be used. It is not suggested that these examples need be followed slavishly as a system in either format or wording; the intent is simply to demonstrate comprehensively each step in the manuscript review process. Indeed, some journals do not acknowledge the receipt of manuscripts that are returned by referees without reviews. Other journals acknowledge the receipt of referees' reports by sending photocopies of the other referee's report and the editor's decision letter. However, the examples may serve as a guide in setting up new editorial offices or to indicate gaps that have been overlooked in established systems.

To Authors—Acknowledgment of Submissions

Authors like to know that their manuscripts have arrived safely in the editors' offices. The following card acknowledges receipt and gives the official date of submission and the manuscript control number for future reference.

Journal Title

Date: (that card is sent)

Dear Author:

The manuscript entitled _____

(sufficient blank space to type in title)

_____

has been received and was sent to referees today.

The Editors

Date received: _____

Control number: _____

To Referees — Initial Letters

Manuscript Control Number: _____

Dear (referee):

   The enclosed manuscript (authors and title given below) has been submitted for possible publication. We would appreciate receiving your comments on this manuscript and your opinion regarding its suitability for publication in this journal. As we are trying to keep the interval between submission and publication to a minimum, we would be grateful if you could return the manuscript with your report by the date specified below. If you are unable to meet this deadline, would you please return the manuscript immediately, or call me by telephone (000-000-0000).

Yours sincerely,
Editors

Please report by: _____

Authors: _____

Title: _____

The submission of two, related manuscripts happens with sufficient frequency that a modified form letter to referees is worthwhile. The previous letter can be amended as follows:

Dear (referee):

   The enclosed two manuscripts (authors and titles given below) have been submitted for possible publication . . . telephone (000-000-0000).

   We apologize for sending two manuscripts at one time but they are closely related and by the same authors so it seems best that they be reviewed together.

Yours sincerely,
Editors

Any requests for referees to review more than two manuscripts at one time should be made by a special letter that sets forth the circumstances.

To Referees — Acknowledgment of Unreviewed Manuscript

   Referees who have been unable to review a paper but return the manuscript promptly may be notified of its safe arrival.

<u>Journal Title</u>

Date: (that card is sent)

Dear Referee:

We have received the manuscript that we sent to you recently for review,

Title: _____

_____

Authors: _____
for which we thank you.

The Editors

To Referees — Reminders

Journals should have a form letter to prompt those referees who have not sent in their reports on time. Such letters are rarely resented if the requested deadline is realistic, and serve to show that the editor's office is alert.

> Dear (referee):
>
> We note that the manuscript listed below, sent to you for review on (date), is still outstanding. We would be grateful if you could examine this manuscript and return it, together with your report, immediately.
>
> Yours sincerely,
> Editors
>
> Authors:
>
> Title:

To Authors — Reports of Delays

When referees are slow, editors can forestall queries and complaints from authors by letting them know what has happened. Most authors understand the vagaries of timing in the review process and will not complain provided they are kept informed and know that their manuscripts have not been forgotten. The following three form letters to authors will cover most situations.

> Dear (author):
>
> This is just to inform you of a modest delay in handling your manuscript, (title). One of the referees is being slow. We have sent a reminder

that we hope will speed things along and shall write to you again when the report has been received.

> Yours sincerely,
> Editors

A plural form of that letter may also be useful.

Dear (author):

This is just . . . Both referees are being slow. We have sent reminders that we hope will speed things along and shall write to you again when the reports have been received.

> Yours sincerely,
> Editors

Then there will be the occasion when a referee was not able to review the manuscript. Again, the authors shold be kept informed by a form letter along the following lines.

Dear (author):

This is just to inform you of a modest delay in handling your manuscript, (title). One of the referees chosen originally was unable to act so the manuscript has been sent to an alternate. We will send a reminder if the report is not in shortly and will write to you again when we have heard from the referee.

> Yours sincerely,
> Editors

To Referees — Acknowledgment of Reports

When referees have been particularly helpful, or when it seems obvious that editors will want them to check revised manuscripts, the editors will probably want to acknowledge the initial reports by special letters. When, however, reports are fairly routine and criticisms are minor, a simple acknowledgment card is a convenient way to let referees know that their reports have been received and are appreciated.

Journal Title

Date: (that card is sent)

Dear Referee:

We have received your report on

Title: _____

_____

Authors: _____

for which we thank you.

The Editors

Possibly a better form of acknowledgment is to send each referee photocopies of the other referee's report and the editor's decision letter to the authors. This procedure is now used by several journals and has been very favorably received.

To Authors — Notification of Acceptance

When authors have met all of the editors' requirements for revision of their papers, there is generally no need for editors to write a specific letter; a simple notice of acceptance is all that is required.

Journal Title

Date: (that card is sent)

Dear Author:

The manuscript entitled _____

_____

has been accepted for publication and will appear in an early issue.

The Editors

The Manuscript Files

The heart of any editorial operation lies in the manuscript files where a copy of each manuscript is kept to protect against loss and where there is a complete history of each paper in the system. Each paper should have its own file folder that should contain a copy of the manuscript, a control sheet, and a copy of every document that arises during the review process. The control sheet is conveniently stapled or glued to the inside of the front cover, and all documents, other than the manuscript, should be fastened into the folder; loose pieces of paper have a tendency to wander.

Next to the manuscript itself, the most important item in the manuscript file is the control sheet because it permits an instant reading of the status of the paper. Control sheets can be set up in many different ways. The important thing is that there are places to note all details about the manuscript and each step in its review. Figure 11 shows one possible arrangement for a hypothetical journal, "Trees." The journal is identified at the top of the sheet where there is also space for the control number assigned to the manuscript. This is the number by which the manuscript is filed and identified in all subsequent steps of the review process. The control number consists of a letter code "T" for the journal, the "83" indicates the year, and "6" shows that this manuscript was the sixth to be submitted in 1983. The manuscript is then classified as an Article, Note, or Communication, and this is followed by the authors' names and the address of the corresponding author as given in the letter of submission. The name of the author who sent in the manuscript should be underlined to indicate the person to whom all correspondence should be sent. Then comes a description of the manuscript including title, number of pages, figures, and tables, and the number of manuscript copies. A note should be made of the number of figures and how many sets of prints were received. Of course, it is important to note the date that the manuscript was received because that date will be published with the paper. The date that the submission was acknowledged simply serves to confirm that the authors were informed that the manuscript was received. The "comments on format" can be used to note any deviations from the Instructions to Authors that are found in a preliminary examination of the manuscript. In the sample shown, it may be assumed that reference 6 was not a proper reference; perhaps it referred to a personal communication or an abstract from a conference. The remaining part of the control sheet deals with the review process. The referees are listed by name and address with spaces for the dates on which the manuscript was sent to them, when the reports were returned and acknowledged, and when reminders were sent, if necessary. Under "comments" in this section, the editor can write a brief description of the nature of the referees' reports, and make a note if the referee should be asked to check the revised paper. Spaces are provided to show when authors have been informed of any delay, and the final section records the disposition of the paper. Following through the example, the whole history of the paper is clear. It was received on January 6th, acknowledged on January 7th, and sent to referees on January 10th. The second referee responded promptly; the report was received January 26th and acknowledged by card the next day. The first referee was slow, possibly because of the mail service, so a reminder was sent (February 4th) and the authors were informed of the delay. The first referee's report finally arrived on February 15th and was acknowledged by a card on February 16th. The manuscript was returned to the authors for revision on February 17th. The editor has noted that referee 1 had only minor comments and that referee 2 had major criticisms but did not need to see the revision. Presumably, the editor felt competent to check that this referee's criticisms had been met. The revised manuscript was received on March 28th and was accepted by the editor on March 30th. An acceptance card was sent the next day.

A control sheet such as shown (Figure 11) allows for a record of all the essential dates and events that occur with a majority of papers. It would probably be counterproductive to try to devise a form that would

Control No.: T.83-6

T R E E S

Article: ☑

Note: ☐

Communication: ☐

Authors/address: Arthur J. Lumber and Peter M. Jack
The Chainsaw Polytechnical Institute
Bushland, Alaska 01010, U.S.A.

Title: The Growth of Trees During Earthquakes

Pages: 10   Date received: Jan. 6/83

Figures: 4 drawings (Figs. 1-4)   Acknowledgment
2 sets of prints   card sent: Jan. 7/83

Tables: 2   Comments   Reference 6 should
                                on format: be a footnote

Copies: 1 typescript
+ 2 photocopies

Referees: (1) Dr. N.O. Shake   Sent: Jan. 10/83   Comments
Dept. of Seismology   Rec'd: Feb. 15/83   minor comments
Allsorts University
Earthquake Island   Ack'd: Feb. 16/83
Greece   (card)

                                        Reminder: Feb. 4/83

(2) Dr. R.U. Forest
Dept. of Treeology
Woods University
Parkland, Maine 02020
U.S.A.

		Comments
Sent:	Jan. 10/83	
Rec'd:	Jan. 26/83	major criticism but no need to see revision
Ack'd:	Jan. 27/83 (card)	
Reminder:		

To authors for revision:	Feb. 17/83	Rec'd:	Mar. 28/83
Withdrawn:			
Rejected:			
Accepted:	Mar. 30/83		
Acceptance card sent:	Mar. 31/83		

Modest delay letter: Feb. 4/83

Further delay:

Figure 11

account for all possible aberrations; written notes in the margins or blank areas of the form can be used to record any unusual events. For example, in the hypothetical history just described, had the editor wanted referee 2 to check the revised manuscript, it would have been so noted under "comments," and the initial referee's report would probably have been acknowledged by letter instead of by card. Then, when the revised manuscript was received from the authors, it would have been sent back to referee 2 and the dates for sent, received, and acknowledged could be recorded under an additional heading "Rev," for revision.

It is convenient to have three copies of the control sheet, color coded to make for easier filing. The top copy is placed on the inside cover of the manuscript file; this is the master sheet that is updated continuously to show the status of the manuscript. The second copy is filed by control number, and the third copy is filed alphabetically by the name of the first author. Any time that the manuscript file is pulled, the master control sheet shows what needs to be done. The file of control sheets by number provides a complete record of the activities of the editorial office and can be used retrospectively for the compilation of various statistics. The alphabetical file serves as a cross reference to locate a manuscript file when incoming correspondence has neglected to refer to the control number. Of course, the multiple filing can be eliminated if the records are stored in a word processor or microcomputer that can be programmed to sort the information in various ways as required.

The manuscript file, labeled and stored by control number, should contain at least one copy of everything having to do with that manuscript. The first items will be the letter of submission and a copy of the manuscript. Some journals do not retain a manuscript copy, but it is useful for the subsequent check of revisions. Then will come copies of the letters to referees, the referees' reports, and any correspondence with the referees. Next there will be a copy of the editor's letter to the authors about the referees' reports and advising on the next step. Similarly, any further correspondence with the authors or referees should all be placed in the file in chronological order until the file is completed by acceptance, rejection, or withdrawal of the manuscript.

Manuscript files require a considerable amount of secure storage space, and accumulation can present problems. Editors need to establish procedures for periodic clean outs of dead files. Certainly, the files on papers that have appeared in print can be discarded, although editors may want to retain them for some predetermined period (6 to 12 months) against the unlikely event that some questions arise. Because of possible resubmissions, a longer holding period may be indicated for papers that have been rejected, withdrawn, or returned for major revision. It is important to remember that, even though a paper has been published, the referees' reports and identities are still confidential information and part of the editors' responsibilities. Files that are discarded should be destroyed.

The Referee Files

The main reason to maintain referee files is to avoid too frequent use of the same referees. Referees will be much more willing to review papers if they know that there are specified intervals between requests. To respect that interval, editors need to have a record of the last date on which a particular referee was sent a manuscript for review. The simplest system is a card index, filed alphabetically, that contains on each card a referee's name, address, and telephone number, followed by two columns: one headed "manuscript sent" and the other "report received." Entry of the appropriate dates on the card then provides a complete reviewing history of that referee. This simple system can be embellished as desired by addition of a column for comments ("good report" or "poor report") or space for notes such as, "should not be sent papers by Dr. J. Blow" or "is slow but worth the wait." A large card will permit the inclusion of the titles and authors of each manuscript that the referee reviewed. The information can be useful in revealing closed cycles, i.e., two referees who are consistently asked to review each other's papers, an indication that fresh viewpoints might be sought.

The Diary

Every editor's office requires a diary or calendar in which should be noted, in advance, the control numbers of manuscript files that need to be brought forward for action on a particular day. The diary is of special importance for follow-up on delinquent referees. For example, assume that a manuscript is sent to referees on January 10th with a request to review within two weeks and that the journal allows a week's grace to account for time in the mails. When the manuscript is sent to referees (i.e., on January 10th), its control number should be entered in the diary page for January 31st. Then, when January 31st arrives, the control number on the page for that day indicates that the manuscript file should be checked for further action. If the referees have not reported, reminders can be sent and authors can be informed of the modest delay. At that stage, a further entry of the control number should be made in the diary for, say, February 14th — another two weeks — for a further follow-up of the referees, by telephone or telex, if the reports have still not arrived. Similar diary entries can be made when manuscripts are sent to authors for revision (follow-up in, say, 6 to 8 weeks) or when referees are asked to check a revised manuscript (follow-up in 2 weeks). The whole objective is to use the diary as an alerting mechanism for actions required to avoid undue delay in the review process.

Of course, the diary is also useful to alert editors to other items of journal business. Notes about meetings with associate editors, publishers, printers, or copy processors can be made far enough in advance for the proper preparation and assembly of documents, reports, and statistics.

The use of a diary, in the prospective sense described here, is the key to an efficient editorial operation by assuring a smooth, reliable, and efficient flow of manuscripts through the review process and by providing advance warning of items on the editor's agenda.

The Office Operation

The previous sections of this chapter dealt with the tools of the editorial office: forms, letters, cards, files, and diaries. Of course, the tools have to be used in an organized fashion by someone, be it the editor or the editor's assistant. When new people take over these duties, as they will from time to time, the transitions will be eased if some guidance is provided to show who does what and when. It is also important to remember that the flow of material into the editorial office does not stop during absences or changes in personnel. To avoid the buildup of a backlog during those periods, it is essential that new personnel be able to pick up the system quickly and easily. For all of these reasons, each journal should have an operations manual that spells out what needs to be done, and who does it, at each stage in the review process. Because different journals use different systems of referee selection and editorial decision making, it is impossible to draw up an operations manual that would be universally applicable. For simplicity, the description of an office operation given here is based on the fully centralized system of an editor who selects referees and does all the correspondence, and an editorial assistant who runs the office. Obviously, a decentralized system requires a different organization of the material: a section for the central secretariat or editor-in-chief's office and subsections for the other editors or coeditors. The essential point is that each step be defined for each of the participants.

OPERATIONS MANUAL

1. *Mail*

Incoming mail should be opened immediately and sorted into two piles: new submissions and other. New submissions should *always* get priority treatment from both the editorial assistant and the editor because it is important to get those manuscripts out to referees as quickly as possible.

2. *New manuscripts*

The editorial assistant assigns a control number and writes or stamps it on all copies of the manuscript, counts the number of pages, tables, and figures to make sure that each copy of the manuscript is complete, and then passes one copy to the editor with a note, "We need referees." While the editor is selecting referees, the editorial assistant completes the control sheet and sends an acknowledgment card. When the editor returns the manuscript with the names and addresses of referees, the editorial assistant checks the referees' file to see if they are available, enters the names, addresses, and date on the control sheet, prepares the form letter for the referees, and sends out the manuscript. If the referees have been used within the interval specified by the journal, the manuscript should be returned to the editor for selection of alternates. The package to each referee should contain: one copy of the manuscript, the letter to the referee requesting a review, copies of the referee report form, and a preaddressed return envelope with sufficient postage to cover the mailing costs. If the manuscript is being sent out of the country, international postage coupons can be used, or foreign referees can be reimbursed by check or money order after their reports have been received.

A copy of the manuscript, the letter of submission, copies of the letters to the referees, and the top copy of the control sheet are then placed in a file folder that is labeled with the control number and filed. The control number should be entered on the appropriate diary page for follow-up on the referees if necessary. The second copy of the control sheet is filed by control number, and the third copy is filed alphabetically by first author.

The date on which the manuscript was sent is entered under the referees' names in the referee file.

3. *Referees' reports*

When referees' reports arrive, they should be placed in the manuscript file, together with any covering letters, and passed to the editor. The editor should indicate immediately whether the report should be acknowledged by a card or a letter. Dates of receipt and acknowledgment are entered on the control sheet and on the referees' file. When reports have been received from both referees, the editor drafts a decision letter which is then typed by the editorial assistant and sent to the authors together with the unsigned copies of the referees' reports and copies of the manuscript for revision. Note: One copy of the manuscript should be retained in the manuscript file as a master copy against which revisions can be checked. At this stage the control number of the manuscript should be entered in the diary, at whatever interval the journal allows for revisions (say, 6 to 8 weeks), so that authors can be queried if they have not sent in a revised manuscript by that time.

4. *Revised manuscripts*

When authors send in manuscripts that have been revised according to the editor's advice, the editorial assistant should (a) enter the date of receipt on the control sheet, (b) check that each copy of the revised manuscript is complete, (c) mark each copy as "revised" (to avoid confusion with the original copy on file), (d) place any covering letter in the manuscript file, and (e) pass the file, with one copy each of the original and revised manuscripts, to the editor for decision. The editor may accept or reject the manuscript, or ask for a further review by one or both of the referees.

If the revised manuscript is accepted, the editorial assistant sends an acceptance card to the author, completes the control sheets, and passes the manuscript to the journal's copy processing unit to be prepared for printing.

If the revised manuscript is rejected, the editor drafts a letter to be typed by the editorial assistant and sent to the authors with copies of the manuscript, referees' reports, and whatever other supporting documents the editor wants included.

If the revised manuscript is to be reviewed further, the editor will probably want to draft a special letter to the referee or arbitrator. The editorial assistant should then type that letter and send it out, with copies of the original and revised manuscripts and a copy of the author's letter of resubmission if appropriate. As always, this information and the date should be entered on the control sheet, and an entry should be made in the diary for a reminder to the referee if necessary.

The foregoing detailed description of the office operation may be supplemented by a checklist (Table 2).

Individual editors can embellish this list as desired for their particular office requirements. Decentralized editorial systems, in which several people may be involved in selection of referees and correspondence with authors, may require some additional paper flow such as extra copies of control sheets, correspondence, and referee files. The important points are that journals should designate clearly the office where authors can find out the status of their manuscripts, and that there be one single office where all information on the journal operation is maintained.

Table 2 Checklist of procedures for editorial assistant and editor

Editorial Assistant	Editor

NEW MANUSCRIPTS

1. Count pages, tables, and figures in each copy	
2. Send acknowledgment of receipt to author	
3. Pass to editor for referees	1. Select referees
4. Check referee file for availability and enter new referees	
5. Complete control sheets	

6. Send out to referees:
  a. form letter
  b. manuscript
  c. referee report forms
  d. return envelope with postage	

7. Make entry in diary of date to remind referees	
8. Enter on control sheets and referee file the dates on which referees' reports are received	
9. Send acknowledgment cards or letters to referees	2. Draft letters to referees if necessary
10. Put referees' reports in manuscript file and pass to editor	3. Draft decision letter: accept, reject, or revise

11. Type editor's decision letter about revision and send to authors with:
  a. all but one copy of manuscript
  b. referees' reports (unsigned copies)
  c. marked Instructions to Authors (if necessary)	

12. Update control sheets	

REVISED MANUSCRIPTS

1. Enter date received on control sheets	
2. Label all copies as "revised"	
3. Add author's letter and revised manuscript to the manuscript file and pass to editor	1. Check revision and decide on accept, reject, or re-review
	2. Draft letters to authors and referees as required

ACCEPTED MANUSCRIPTS

1. Complete control sheets	
2. Send acceptance card or letter to author	
3. Pass manuscripts for copy processing	

Table 2 (continued)

Editorial Assistant	Editor

REJECTED MANUSCRIPTS

1. Type editor's decision letter
2. Send letter, copies of manuscript, and
   referees' reports to author
3. Complete control sheets

FURTHER REVIEW OF REVISED MANUSCRIPTS

1. Type and send editor's letter to authors
2. Type editor's letter to referees and send with
   (as directed by editor):
   a. copies of original and revised manuscripts
   b. referee's forms
   c. copies of referees' reports on original
       manuscript
   d. copy of authors' letter of resubmission
   e. return envelope and postage
3. Enter information on control sheets
4. Make diary entry for follow-up on referees
5. Update control sheets when report is received
6. Acknowledge report by card or letter
7. Pass file to editor for decision

    1. Draft letter to referees
    2. Draft letter to authors:
       accept, reject, further
       revision

8. Type editor's letters and proceed as directed
9. Keep control sheet and diary up to date

The Journal's Data Base

The extra copies of the manuscript control sheets that are filed by control number constitute the data base of the journal from which various kinds of information can be obtained.

The last control number for the year represents the total number of submissions. A significant decrease in that number from year to year might signal a drop in popularity of the journal that is worthy of investigation. The percentages of rejections, acceptances, and withdrawals can be calculated from the numbers of those categories on the control sheets. The results may be useful in establishing whether editorial policies are too tough, too lenient, or moving in the right direction.

The authors' addresses on the control sheets permit an analysis of the sources of papers that may be helpful in answering such questions as: How international is the journal's authorship? Are there strong laboratories or groups that are not represented in the journal? Is the journal getting a fair share of papers from specific laboratories in universities, government, and industry?

The dates on the control sheet are important measures of the time taken for the various steps in publication of a paper. An analysis may indicate areas that can be improved. For example, if there is a difference of a week or more between the dates when submissions are received and the dates on which they are sent to referees, then it is clear that new manuscripts are not being given the priority treatment that they deserve; the difference in those dates should not exceed one day. If referees are slow to report, perhaps the follow-up procedure is not being used correctly. On the other hand, if a significant number of referees are consistently late by just a few days, then the journal's deadline may be unrealistically short. Editors should remember that authors compare journals for speed of publication by noting only the difference between "date received," marked on each paper printed, and the date of issue of the journal. A lot goes on between those two dates, and the editors can influence most of it by riding herd on referees and by pressing the copy processors and printers when undue delays occur. The period least amenable to editorial control is the time taken by authors to revise their manuscripts. If analysis of that time interval, shown on the control sheets, shows that the average revision takes more than, say, six to eight weeks, then editors might consider two possibilities: (1) institute a follow-up procedure for authors similar to that for referees, or (2) establish a deadline beyond which a manuscript will be regarded as a new submission and sent out for review or marked with the date, "revision received," when printed.

Finally, the total number of manuscript pages, figures, and tables, obtained from the control sheets, can be useful information for the negotiation of printing contracts and for assessing the copy-processing capacity required by the journal.

One cautionary note may be made about the use of statistics in editing a journal. There is a natural tendency to regard numbers as absolute or "hard." However, in the journal business much of the data is "soft" because there is so much variability. For example, an "average" time interval for refereeing is not very meaningful in itself because of extremes; three or four quick reports can be offset by only one that is slow. A better analysis of referees' times is given by a bar graph that shows the percent of reports received after two, three, four, . . . weeks. The journal's data base is better used to indicate trends rather than in any absolute sense. A record of various journal data, accumulated in the same way over a period of years, can be a valid indicator, and corrective action can be taken if an undesirable trend is detected before it becomes a major or irreversible problem.

Automation

Many editors are already using microcomputers or word processors in their editorial offices. Just about everything mentioned in this chapter can be handled more easily and efficiently on a microcomputer with word-processing and printout capabilities. Starting with the manuscript control sheet, the equipment can be programmed to print all of the information in the appropriate places to provide a hard copy and, of course, the same information can be stored on a disc or tape for recall at the press of a button. Similarly, a journal's entire file on referees can be stored in a microcomputer much more conveniently than in a card index. The microcomputer can also be used in the follow-up process by including an "alert" file that contains manuscript control numbers and the dates for reminders to referees, letters to authors, or any other date-dependent information.

Word processors are invaluable in dealing with routine form letters, acknowledgments, cards, and envelopes. Almost all equipment can be programmed to print any information in its proper place on a form letter, card, or envelope. Thus, it is only necessary to enter the new information—names, addresses, titles, dates—and the processor will print it in the designated spaces on the form letter, card, or envelope. Most equipment can also be programmed to sort or order files on the basis of names, addresses, dates, titles, or key words with obvious advantages for cross-referencing or compilation of indexes.

The capabilities of electronic processing as just described are here and now, ready for application in any editor's office. The only other considerations are those of expense and volume of work. On the latter point, any journal that has a daily flow of paper sufficient to keep a full-time editorial assistant fully occupied at a typewriter could probably benefit from word-processing equipment. The expense is determined largely by the capabilities of the equipment, and there is a wide range with a hundred or so word processors on the market. To make a choice, editors and their editorial assistants should define clearly their needs and objectives, and consult extensively with equipment manufacturers, other editors, and major publishers who have experience in electronic processing. As long as the electronic processing is confined to the routine of the editorial office, the equipment need not be compatible with other models.

Chapter 8

Copy Processing and Printing

Who Does It?

Scientific editors normally should not be expected to participate directly in copy editing, proofreading, or negotiations with printers. Copy processing, the preparation of manuscripts for printing and the correction of proofs, is a skillful, detailed activity that requires, and deserves, more time than a scientific editor could afford. Publishers should therefore provide this service either by hiring in-house copy editors or by contracting the work out to experienced people. Some printers may offer copy processing as an option in their contracts. For small journals, part-time copy editing can often be obtained through printers or by contracts with individuals. Generally, copy editors are more efficient and accurate if they understand the text being processed; some knowledge of the science covered by the journal is therefore an advantage. A scientific background also enables a copy editor to query an author directly about any changes in the manuscript and thus avoid needless requests to the editor to intercede.

Negotiations with printers should be done by a managing editor or a business manager who is knowledgeable about the printing business and the requirements for scientific journals. Such knowledge is built up and maintained in the journal's business office, which should accumulate information on printing specifications, prices, and the capabilities of various printers. Specifications in printing contracts are highly detailed, require careful preparation, and should be reviewed from time to time if for no other reason than to see whether economies can be achieved. The editor need not be involved in any of this work except, as noted below, when major changes are contemplated.

The Editors' Interests

The quality of any journal is based on its contents and its physical appearance, and there are some obvious connections between the two. Editors who want to improve the contents of their journals will try to attract papers written by the best scientists in the field. Those scientists will not want to publish in journals where there is a high risk of typographical errors, where the print is sometimes smudged, small, or difficult to read, where carefully prepared photographs have lost all their detail, and where the paper used for printing resembles either onion skin or newsprint. These are exaggerations, of course, as no journal could survive with all those faults. However, they do serve to indicate why editors should be concerned with the production side of their journal and the need for editors to maintain an active liaison with the people involved.

The quality of the physical appearance of a journal depends directly on the funds available for its production. Thus, the standard of production will usually be a compromise between what is acceptable and what the publisher can afford. Editors, and their editorial boards, can assist by defining what is minimally acceptable, what is desirable, and what would be best, for their journals. The standards of production involve the level and extent of copy editing, the format of papers, the format of the journal, type style and size, quality and weight of paper, quality of half-tone reproductions, binding, and cover design. Editors and members of editorial boards can consult colleagues in their home institutions and thus provide extensive and valuable feedback on each of those items. The objective should be to publish a journal that is free from errors, attractive, and easy to read, and in which the quality of reproduction is consistent with the scientific content.

Finally, and perhaps most important, editors will be concerned that their journals appear on time. Speed of publication is a high priority for authors, so editors should try to keep their journals competitive with others in the same field. Editors themselves can try to speed up the review process, but they should also be interested in the efficiency of the copy editing and printing and should not hesitate to question any slowdowns. There will always be the occasional, unexpected delay caused by a turnover in staff, a press breakdown, or labor problems. However, delays should be only occasional, and any consistent failure to meet schedules should cause editors to urge corrective action by the managing editor or business manager.

How Much Copy Editing?

The term copy editing means different things to different people, and editors should have a clear understanding with their managing editor and copy editors about what they expect from each other. Editors are concerned primarily with the scientific content of the papers that come to them and may pass manuscripts for printing without paying any attention to the format or the state of the copy. The result may well be the introduction of errors, delays because of uncertainty, and the generation of antagonisms and misunderstandings. If a paper is not in the stipulated format of the journal, the copy editor will have to restructure it, which takes a lot of work and runs the risk of distorting the author's meaning. If the copy contains a lot of hand-written revisions, the copy editor will have to either retype those parts or gamble that the printer's compositor can read it. Either route wastes time and increases the potential for errors. Editors should not accept papers until the manuscripts are ready for the copy editors. Authors, being the most knowledgeable about their papers, are in the best position to make any revisions or changes in format without the introduction of errors or distortion of meaning. Editors can save time for all concerned by including requests for changes in format with their letters about revisions to the manuscripts on the basis of referees' reports. It is a simple matter to enclose a copy of the Instructions to Authors, suitably marked in areas where the paper is deficient, and to request authors to deal with those points in revising their paper.

If manuscripts passed for printing are all in proper format and clean copy, what is left for copy editors to do? The instructions to the printer are probably of prime importance. That involves careful marking of the manuscript in accord with the editorial style of the journal, including clear directions about type fonts and sizes for all parts of the manuscript: title, headings, text, tables, legends, footnotes, and references. There is also a need to indicate clearly any italics, boldface, Greek letters, and mathematical symbols. Then the copy editor should check nomenclature, abbreviations, units, and symbols to see that they are consistent and in agreement with whatever internationally accepted system the journal has adopted. After that comes a check of tables and figures to see that the results are presented clearly, that all are referred to in the proper place in the text, and that titles and legends are correct. Finally, copy editors should check that the manuscript is complete, that all parts are in order for the printers, and that all references are cited and in the form used by the journal. Copy editors for journals are expected to correct any errors in spelling or grammar, and are sometimes encouraged to make minor revisions in the text where clarification seems required. This latter point can become contentious with authors and must be handled carefully and judiciously to avoid antagonism. All papers have been read by referees who, while not expected to do any extensive rewriting, should certainly note any ambiguities or unclear passages. Thus, copy editors should be certain that their proposed changes are actually required for clarification and are not just a matter of style. At the very least, authors should be given the chance to disagree, either by a telephone call when the copy editing is being done, or by a discreet query on the manuscript that is sent to the authors with proofs. Failure to take such precautions can result in letters from authors like the following:

> Dear Copy Editor:
>
> I fail to understand the continuing tendency of copy editors to rephrase manuscripts with which they are presented. The original phrasing is mine (as it should be in any original paper) and I wanted it to be mine, not that of some unidentified copy reader who would rather have it another way and so writes in his or her own "style."
>
> I strongly object to changes made for the mere sake of change. If it is grammatically correct in the original (even if it is not pleasing to the editor), the manuscript should not be altered. An outstanding example in this manuscript is the change of "while" to "whereas" throughout. May I refer you to Fowler (p. 722 in my volume) to support my contention that the original was entirely satisfactory.
>
> I appreciate the normal efforts of any copy editor to make a manuscript intelligible to the typesetter. The appearance of the galleys for this manuscript show that this job has been done with a high degree of skill, but I think that the effort should end there. Alterations in phrasing should not be permitted especially when it is a completely one-way proposition, i.e., the author sees these only after the deed is done.

Probably a majority of authors feel the same about copy editing as the writer of that letter. For that reason, editors and copy editors should be clear about how much revision should be done to a manuscript that has been passed for printing. A useful and time-saving guide is to limit revisions by copy editors to correction of errors; rephrasing should only be contemplated when there are obvious ambiguities or unclear expressions. What matters is the scientific content, and as long as that is expressed clearly in a retrievable form, the literary style is of lesser consequence.

A special problem arises with papers written in a language with which the authors are not really familiar. With the current predominance of English as the language of science, many authors whose mother tongue is not English want to publish in the language that is read by the majority of scientists. What does an editor do when faced with a paper that reports good science according to referees, but is filled with grammatical and spelling errors, and many examples of improper use of language? Several options are open, the choice depending upon available time and the basic humanity or goodwill of editors, referees, and copy editors. Of course, if a paper is so poorly written that the science cannot even be understood, the editor is justified in returning it directly to the authors with the suggestion that they seek assistance from someone more familiar with the language. If papers are understandable, even though poorly written, editors can send them to referees to see what happens. Referees are not expected or required to make extensive revisions in the text of papers sent to them for review. However, some referees, in a display of truly enormous goodwill towards their international colleagues, will undertake detailed revisions of papers where there are obvious linguistic problems, particularly if those papers report good science. Other referees may be enthusiastic about the science in a paper and sympathetic towards the language difficulty but simply not have enough time to rewrite the text. When that happens, editors can suggest that authors seek assistance from someone knowledgeable in both the language and the science, maybe even suggesting the names of scientists who might be approached, or offer the services of the journal's copy editors. This latter option is clearly a tricky one because extensive text revision of even a few papers may slow down the copy editing and introduce delays in the publishing schedule. The decision is up to individual editors who can best judge whether or not a paper that is deficient linguistically warrants the extra copy editing and whether or not the journal's copy editing capacity can handle it.

In these days of hustle and bustle it is perhaps not surprising that the whole concept of copy editing has been questioned as being an outdated waste of time and money, particularly time. However, suggestions to eliminate copy editing overlook the fact that printers' compositors generally are not trained in science and function solely on the basis of directions given in the manuscripts. With a lot of pressure from editors there might be a few authors who would provide pristine, clear typescripts, but not many would be prepared to mark the manuscript for printing or do the thorough checks outlined earlier in this section. Elimination of copy editing would therefore lead to a decline in standards of publication and increases in corrections and alterations of proofs for any papers that are typeset or rekeyboarded before printing.

Some journals have reduced the amount of formal copy editing by the use of "camera-ready" typescript provided by authors. However, in those instances manuscripts must be prepared according to detailed specifications, sometimes typed on special paper provided by the journal, and proofread carefully to catch any errors. In other words, a major portion of the copy editing function has been assigned to the authors and their typists. The results will obviously depend upon the care that authors and typists are prepared to devote to manuscript preparation; standards are therefore likely to be highly variable. A procedure has been evolved that should reduce this variability in standards. The final version of the manuscript is copy edited by the publisher and then returned to the author with specific instructions for the preparation of camera-ready copy. Figure

Preparation of Final Author-Produced Copy for JGR

To the Author

The format set forth in these instructions must be followed exactly. Copy that does not meet JGR specifications will be returned for retyping. In particular note the following items:

● Review the edited manuscript and answer all questions directed to the author before you give the paper to your typist. Make no changes of a technical or substantive nature without obtaining the Editor's approval.

● Final copy must not exceed the specified widths: 10.4 cm (4 1/8 in.) for single column, 21.6 cm (8½ in.) for double column, 31.7 cm (12½ in.) for broadside tables or figures.

● Unless IBM's Prestige Elite 72 and Symbol 12 Selectric elements are used, you must obtain prior written approval for the typewriter face that will be used. Use of an unapproved typewriter face may necessitate retyping the entire paper.

● Mathematics must be typed. Your typist must follow the instructions below for handling characters not available on a standard typewriter.

● Check the final copy carefully for errors; it will not be proofread at AGU. The typescript will be photographed exactly as it is received; insuring the accuracy of the final typing is your responsibility. Complete the Author's Checklist as you make your final check.

● Refer to the letter accompanying your manuscript for the correct percent of reduction for your line drawings. (Continuous tone photographs or original figures of correct size have not been returned to you.) Figures plus their legends must not exceed the dimensions of the type page. All figures will be printed at 80% of the size of the final copy; therefore no lettering should be less than 2 mm high. Submit original ink drawings or sharp black and white glossy prints. If you are unable to provide figures at the size recommended by the copy editor, AGU will size them for a charge of $10 per figure.

● If you cannot return the final copy within 1 month of receipt of these instructions, please notify the AGU editorial office of the date you expect to return it.

● Ask your typist to read through the detailed instructions before typing is begun. A sample typescript is included for information; however, your typist should feel free to call the AGU office for clarification of any detail.

To the Typist

Before you begin typing, read through all of the instructions and study the accompanying sample. Failure to follow these instructions may necessitate retyping the paper. If you have any questions, feel free to call the AGU editorial office (202) 462-6903.

6-3-81

● Under no circumstance may you exceed the dimensions of the type page: 10.4 cm (4 1/8 in.) wide for single column, 21.6 cm (8½ in.) wide for double column, 31.7 cm (12½ in.) wide and 21.6 cm deep (8½ in.) for broadside tables.

● Be certain to use the approved typewriter face: IBM's Prestige Elite 72 or the face that has been approved by AGU in writing.

● Use a nonporous good quality white bond and a fresh ribbon. Paper from which typing can easily be erased, tissue paper, and photocopies are unacceptable for reproduction. A carbon ribbon provides the best photographic copy. Please be sure that the type element is clean so that a sharp black image is produced.

● Keep the copy clean. For correcting errors, we recommend correction tape. 'White out' or its equivalent may be used, but care must be taken to insure that it does not flake off. Do not place cellophane tape over any of the type.

● Type all material single-spaced. Refer to the sample for spacing around headings. Do not leave extra space between paragraphs.

● Number each page of the typescript in the upper right-hand corner.

● Label each illustration on the back in light blue pencil with the author's name, figure number, and an arrow pointing to the top of the figure if this is not obvious.

● Mail final copy and figures flat. Use cardboard protection to prevent damage in transit. Include in the package the final typed original, the figures, the edited manuscript, the completed Author's Checklist, and a photocopy of the entire paper including figures and tables. Send by first class mail to AGU, Publications Office, 2000 Florida Ave., N.W., Washington, D.C. 20009. Also send a photocopy of the final typescript to the Editor's office. Check a recent issue of JGR for the address.

Handling Special Material

Mathematics Mathematics must be typed. Only letters and symbols not available on a standard typewriter may be drawn by hand or reproduced from 'lettra-set' or similar lettering tools. Handwritten characters must be clearly drawn with black (preferably India) ink. Leave extra space above and below displayed equations and number them consecutively on the right.

Tables In editing the manuscript the copy editor has suggested table layouts that will be compact and easy to read and has indicated the maximum width the table should be typed and the proper placement of rules. Type all parts of the table single-spaced; double spacing may be used to separate parts of the table or to show grouping of data.

Figure Legends Type each legend single-spaced the width noted on the edited manuscript. Leave extra space between legends to facilitate cutting them apart for paste up.

Figure 12

12 shows an example of such instructions and the level of detail that is required. In addition, the copy editor will have given specific directions in the form of marginal notes on the manuscript as shown in Figure 13.

Typescript consumes more space than typeset text, and the latter is generally regarded as easier to read. Nevertheless, quite a few journals are published successfully using authors' "camera-ready" copy, and more can be expected to follow that route as the technology of text processing improves and becomes cheaper. Publishers and editors will have to determine what level of standards and aesthetic quality will be acceptable to the authors and readers of their journals. "Camera-ready" typescript will probably be phased out in the future as more authors submit floppy disks of their manuscripts suitable for copy editing and resetting.

Relations with Printers

Most of the interactions with printers will be through the copy editors and the managing editors or business managers of journals. Editors who may not be blessed with such assistance will have to educate themselves in the business by visits to printing shops, reviews of sample specifications, and consultations with publishers of other journals. Although the day-to-day contacts with printers will be done by copy editors, the scientific editors should be interested and involved in some aspects of the production process. Probably of prime concern to editors will be the printers' specific ability to deal with certain kinds of scientific material. For example, the ability to set complex mathematics, chemical equations, and structural formulae; the skill and equipment to produce high-quality halftones from photographic prints. A good way to test a new printer's capabilities is to request some trial runs at printing some old manuscripts that fairly represent the most difficult material that has appeared in the journal. Comparisons can then be made with the previously published papers to see if the new printer's work is up to the standard required; scientific editors should definitely be involved in making these comparisons.

The next point of concern to editors will be the printers' ability to produce the journal on schedule. Editors should satisfy themselves that printers have an appreciation of the priority given to speedy publication in science. Nothing does more harm to a journal's reputation than consistently late publication. There should be a clear understanding that the printing schedule is an important part of the specifications in the printing contract and that failure to meet deadlines will be just cause for termination. Site visits can be useful to see if a printer has enough trained people and the press capacity to meet realistic deadlines. If the printer also produces other periodicals, then they can be checked to see if schedules have been met.

middle-level subheadings: cap and lower case, underline, flush left, double-space above and below

Sinusoidal Response

Walsh [1969] has solved for the mechanical response to sinusoidal oscillations of such a system consisting of flat ellipsoidal, or 'penny-shaped,' viscous pockets within an otherwise elastic solid. His results yield for the composite an effective bulk modulus approximately equal to the bulk modulus of the solid. However, the effective rigidity is dramatically reduced by the viscous inclusions and is strongly dependent on viscosity and frequency.

do not leave extra space between paragraphs

In a manner analogous to that of Walsh we solved for the quasi-static relaxation of the same system. The effective shear modulus of an elastic matrix with elastic penny-shaped inclusions is given by Walsh [1969, equation 16]. The shear stress σ and the shear strain ε of the elastic inclusion are related by

equations: center on line, double-space above and below

$$\sigma = 2\mu\varepsilon \qquad (2)$$

equation numbers: flush right in parentheses

where μ is the elastic shear modulus.

low-level subheadings: cap and lower case, underline, with period

Acknowledgments. This study was supported by grant GA 36135X from the Earth Sciences Division, National Science Foundation. G. Mavko was supported in part by a National Science Foundation graduate fellowship.

double-space between text and acknowledgments

References

Billings, M. P., Structural Geology, 3rd ed., Prentice-Hall, Englewood Cliffs, N. J., 1972.

references: first line flush left; indent runover lines 2 spaces

underline title of book, periodical, or report

Morris, E. C., Olympus Mons, paper presented at the Third Colloquium on Mars, Lunar and Planet. Inst., Pasadena, Calif., Aug. 31-Sept. 2, 1981.

Walsh, J. B., Attenuation in partially melted rock, J. Geophys. Res., 74, 4333-4337, 1969.

single-space between references

figure legends: single-space, the width noted on edited manuscript

Fig. 1. Regional flow. Abrupt thrusting of an earthquake causes regions of compression C and dilation D.

TABLE 1. Relaxation With Melt Phase

type line below title

type line between column heads and body

type line at bottom of table (before footnotes)

	Mechanism		
	Viscous Cracks	Regional Flow	Melt Squirt
Relaxation time	$10^{-2}-10^3$s	$>10^3$ yr	3-5 yr
Conclusion	too fast	to slow	reasonable

tables: type each on separate page, single-column (10.4 cm), double-column (21.6 cm), or broadside (31.7 cm)

figures: label each in light blue pencil on reverse, single-column (10.4 cm), double-column (21.6 cm), or broadside (31.7 cm)

Figure 13

Finally, editors should be interested in such things as paper quality, binding, cover design, format, and size. Decisions on each of these items require consultations with printers because only they can provide information on costs and feasibility. Editors, because they are identified on the mastheads as being in charge of their journals, will receive any negative feedback from readers. Thus, complaints that the journal is hard to read, either because of reflections from a high gloss paper or because of the amount of "show through" in a light-weight paper, would be good reasons to discuss the paper stock used in the journal. Complaints that pages were falling out of the last three issues would justify a query about the binding process. Some new editors feel the urge to freshen up their journals with new cover designs and changes in format. The best advice on that is to move slowly; after all, a journal's reputation is based on its content, not its cover. If a change is deemed necessary, the maximum amount of information and feedback should be obtained before making an irreversible decision. On cover designs, the involvement of printers is essential for cost estimates and assistance with art work. The printers' commercial artists can, after discussions with editors and copy editors, make mock-ups of several designs from which a choice can be made. Apart from matters of individual taste, editors should look for a cover design that will make their journal visible and recognizable. Some ideas can be obtained by scanning a display of journals, such as found in the current reading section of a library, to see which journals stand out from the others.

On matters of format and size, editors can provide feedback from editorial boards and readers, but the printers' input is needed for cost estimates and feasibility. As a general guideline, any changes in format should be towards simplification and greater flexibility. A simple and standardized format for positions and type fonts of the various parts of a manuscript allows the copy editors and printers to develop a style sheet for the journal and thereby reduce the amount of mark-up required for the printer. The amount of flexibility required will be determined by the kind of material being published. For example, in journals that carry significant numbers of line drawings, halftones, mathematics, or chemical formulae, a two-column format permits the option of using a full page or half page width for the nontextual material; figures or illustrations can occupy one-half or one-quarter of a page. The page size of journals is almost entirely a matter of economics, determined largely by the printers' press requirements.

Instructions to Authors

The purpose of Instructions to Authors is to speed up the publication process by telling authors how to prepare manuscripts in a form that requires the minimum of revision before printing. Instructions to Authors are not amenable to entertaining writing (although it would be interesting to see some attempts) and there is as yet no universal standard; each journal will have some specific requirements that differ from a general format, and each must therefore publish its own Instructions. Probably the most useful function of Instructions to Authors is as a labor saving device for editors who can use suitably marked reprints of Instructions to point out deficiencies to authors. It is a simple matter for the editor, or the editor's assistant, to note where a manuscript does not conform, underline or highlight the appropriate sections in a reprint of the Instructions, and enclose it with the manuscript when it is returned to the authors with the referees' comments. Editors need then make only a brief addition to their covering letters, such as: "I enclose a copy of our Instructions to Authors with certain points marked for your attention." This is just a polite way of saying, "We have told you this already, if only you would read."

The detailed specification for preparation of manuscripts will be of concern mainly to copy editors who must process the results and who represent the interface with the printers. However, scientific input is essential for several items, so editors need to be involved. Editors and copy editors should review their Instructions to Authors from time to time as requirements change or deficiencies are noted.

Figures 14–16 at the end of this chapter show examples of Instructions to Authors from journals in the fields of chemistry, earth science, and biology. Each of them deals with the following points, not necessarily with equal emphasis or in the order listed.

- *Editorial scope:* A brief statement of the subject matter that falls within the purview of the journal. This is clearly a responsibility of editors and their editorial boards.
- *Classes of manuscripts:* A description of categories such as Articles, Notes, Letters, Communications, Reviews, with any limitations that apply. This is an item of publishing policy that should be dealt with by editors and publishers.
- *The review process:* An account, in varying detail, of the review process used by the journal, including a time limit for authors' revisions beyond which a manuscript is regarded as withdrawn. This item should also be the responsibility of editors.
- *Manuscript preparation:* A guide to the typing and assembly of the manuscript—size of paper, spacing, margins, placement of figure legends and tables, and number of copies required.
- *Title page:* Directions for preparation of the title page, usually to indicate that it should contain only the title, the authors' names and addresses, and any footnotes that are necessary.
- *Abstract:* A note that abstracts are required, the limit on length in

number of words, and a reminder that abstracts should be suitable for indexing and for separate publication in abstracts journals.

- *References:* A description of the system used for citation of bibliographic references in the journal and a definition of what constitutes an unacceptable reference. It is essential here to spell out in detail how references are to be cited in the text (by name or by number) and listed in the bibliography (alphabetically or in order of occurrence), and whether or not titles of cited articles and complete pagination are to be included. It is helpful to give at least two examples to indicate punctuation and the order of authors' names, journal name or book title, title of article if used, date, and pagination.
- *Footnotes:* An explanation of how footnotes are to be indicated and where they should be placed. The symbols or numbers used to indicate footnotes should be clearly distinguishable from those used for references, and footnotes should be placed at the bottom of the page where reference to them is made.
- *Tables:* Specifications that may include the need for a title, a number for reference in the text, avoidance of rules, and how to indicate footnotes to the table.
- *Illustrations:* This item should cover the preparation of line drawings — black ink, type of paper, any limits on dimension, size of letters, allowances for reduction — and the need for original drawings if the journal requires them, with copies for reviewers. Directions may also be given here for the presentation of photographs — glossy, high-contrast prints, how they should be trimmed and mounted, limitations on size, the importance of identification and location in the text. If the journal publishes color illustrations, that can also be mentioned here together with whatever conditions may apply.
- *Spelling, abbreviations, symbols, and units:* The journal should state which standards it uses for each of these categories, name the reference dictionaries, list of abbreviations, or references to units and symbols. Requirements for metric or Système international d'unités equivalents should be specified as well as the need for explanatory marginal notes for any unusual or Greek symbols.
- *Nomenclature:* Most areas of science have developed standards of nomenclature, usually through the International Scientific Unions. Journals should cite the most recently published versions of rules of nomenclature that are applicable, and indicate where they can be obtained.
- *Supplementary material:* Many journals now have facilities for the deposition of supplementary material which can then be made available on demand as a photocopy, microfiche, or microform. Such depositories are intended for extensive supporting material that may be essential for specialists but not needed for a general understanding of

the papers to which they are connected. Details about the submission of material to depositories and in what form it can be made available should be given in the Instructions to Authors.

- *Proofs and reprints:* Authors should be instructed on the method used by the journal for dealing with galley proofs, including any limitations on corrections, charges for alterations, and any time limits for return. Because galleys and reprints both come from the printers, this section may also contain directions for the ordering of reprints.
- *Special items:* Journals in specific areas of science have some special requirements that editors and copy editors may wish to include in their Instructions to Authors. For example, chemical journals may need guidelines for the presentation of structural formulae, chemical equations, and crystallographic data; journals that carry any amount of mathematics will need to give detailed directions for the presentation of mathematical expressions; journals in the biomedical field that subscribe to ethical guidelines on the care and use of experimental animals should so state and include a reference to whatever guidelines are applicable.

NOTICE TO AUTHORS

I. General Considerations

The Journal of Physical Chemistry is devoted to reporting **new** and **original** experimental and theoretical **basic** research of interest to physical chemists and chemical physicists. Manuscripts that are essentially reporting data, applications of data, or reviews of the literature are, in general, not suitable for publication in *The Journal of Physical Chemistry*.

All manuscripts are subject to critical review. It is to be understood that the final decision relating to a manuscript's suitability rests solely with the editors.

Special arrangements with the editor can sometimes be made in order to publish symposium papers as a group.

II. Types of Manuscripts

The Journal of Physical Chemistry publishes five types of papers: *Letters, Articles, Invited Articles, Comments,* and *Errata*.

A. *Letters* are short articles that report results whose immediate availability to the scientific community is deemed important. *Letters* are restricted to 2000 words or the equivalent (~8 double-spaced typewritten pages of text and 3–4 figures). A brief abstract of less than 100 words should be included. *Letters* often will be complete publications, but follow-up publication may occasionally be justified when the research is continued and a more complete account of the work is deemed necessary. No page charges are assessed for manuscripts accepted as *Letters*.

Special efforts will be made to expedite the reviewing and the publication of *Letters*. The time for proofreading the galley proofs is relatively short. For this reason, authors of *Letters* should ensure that manuscripts are in final, error-free form when submitted.

B. *Articles* should cover their subjects with thoroughness, clarity, and completeness but should be as **concise** as possible. Abstracts to *Articles* are limited to 300 words and should summarize the significant results and conclusions. *Articles* are published in the order of acceptance within limitations of available space.

C. *Invited Articles*: A few *Articles* will appear as a result of an invitation from the editorial board and will be so designated. These articles normally will be in active research fields. The author will be asked to provide a clear, concise, and critical status report of the field in a few Journal pages as an introduction to the article. The author's own contribution and its relationship to other work in the field may be discussed clearly. Controversies, if they exist, should also be outlined. Possible future directions and the significance of the research area to the field of physical chemistry or chemical physics should be pointed out. *Invited Articles* are similar to those written for *Accounts of Chemical Research* except that they are written for physical chemists rather than for the general chemical community. Thus, they can be more quantitative, but still should be clear to physical chemists and chemical physicists not directly working in the field. *Invited Articles* are limited to approximately 8 printed pages (equivalent to ~36 double-spaced typewritten pages of manuscript, including figures, tables, and references).

D. *Comments* include significant remarks on work previously published and are restricted to approximately one page (1000 words or equivalent) including tables, figures, text, and a brief abstract. *Comments* are subject to critical review. If the *Comments* are concerned with the work of other authors, the editors will generally permit these authors to reply.

E. *Errata* are author's corrections to printed manuscripts and are limited to the length required to correct errors in print.

III. Submission and Preparation of Manuscripts

The following material should be submitted to M. A. El-Sayed, Editor, The Journal of Physical Chemistry, Department of Chemistry, University of California, Los Angeles, California 90024: (a) three copies of the manuscript and pertinent material that might be required in the reviewing process, (b) a signed copyright transfer form (a copy reproduced from the form printed in the first January issue of the Journal could be used), and (c) a covering letter which specifies the section of the journal for which the work is to be considered. In this letter the author is strongly encouraged to submit names and addresses of three scientists who are competent to review the work critically and objectively.

Authors should include an introductory statement outlining the scientific motivation for the research. The statement should clearly specify the questions for which answers are sought as well as the connection of the present work with previous and current work in the field. In both *Letters* and *Articles*, the introduction should be a separate section of the paper.

In the discussion section, the author should discuss the significance of his observations, measurements, or computations. He should also point out how they contribute to the scientific objectives indicated in the introduction.

Manuscripts must be typewritten, double-spaced copy (one-side only) on 22 × 28 cm or A4 paper. Authors should be certain that copies of the manuscript are clearly reproduced and readable. **Authors submitting figures must include the original drawings or photographs thereof, plus three copies for review purposes. The figures and reproductions should be on 22 × 28 cm paper.** Graphs must be in black ink on white or blue paper. Lettering should be done with a mechanical lettering set or its equivalent. It should be of sufficient size so that after photoreduction to a single-column width (~8 cm) the smallest letter will be 2 mm. Figures and tables should be held to a minimum consistent with adequate presentation of information.

References and explanatory notes should be grouped at the end of the manuscript and typed double spaced. They should be numbered consecutively in the order in which they are first mentioned in the text. Papers should not depend for their usefulness on unpublished material, and excessive reference to material in press is discouraged. Nomenclature should conform to that recommended by the International Union of Pure and Applied Chemistry, the American Chemical Society Committee on Nomenclature, and the Chemical Abstracts Service. For nomenclature advice, consult Dr. Kurt L. Loening, Director of Nomenclature, Chemical Abstracts

Figure 14 Instructions to Authors of *The Journal of Physical Chemistry.*

Service, P.O. Box 3012, Columbus, OH 43210 (tel: (614) 421-6940). Complicated chemical equations, schemes, and structures should be supplied as furnished artwork, ready for photoreproduction. Mathematical expressions and chemical formulas should be typed, with unavailable symbols and letters clearly drawn in ink. Capital, lower case, and Greek letters should be easily discernible, and identified in the margin when ambiguity might result. Avoid complicated superscripts and subscripts. Use fractional exponents instead of root signs.

Consult a current copy of the journal and the "Handbook for Authors of Papers in American Chemical Society Publications" (American Chemical Society: Washington, D.C., 1978) for specific examples of style and general recommendations.

IV. Microform Material

From time to time manuscripts involve extensive tables, graphs, spectra, mathematical derivations, expanded discussions of peripheral points, or other material which, though essential to the specialized reader who needs all the data or all the detail, does not help and often hinders the effective presentation of the work being reported. Such **microfilm material** can be included in the *microfilm* edition of the Journal, available in many scholarly libraries, and also in the *microfiche* edition. In some instances the microform material may also be included in the printed issue as *miniprint*, in which the manuscript pages are reproduced directly in reduced size. All microform material may be obtained directly by the interested reader at nominal cost, either in full-size photocopy or in microfiche (in which miniprint material appears at standard reduction, i.e., one manuscript page per microfiche frame). Authors are encouraged to make use of this resource, in the interest of shorter articles (which mean more rapid publication) and clearer, more readable presentation.

Microform material should accompany a manuscript at the time of its original submission to the editor. It should be clipped together and attached at the end of the manuscript, along with a slip of paper clearly indicating the material is "microform material". Copy for microform material should preferably be on 22 × 28 cm sheets, and in no case on sheets larger than 28 × 43 cm; if typed it should be one and one-half spaced, and in any event the smallest character should be at least 1.5 mm in size (2.5 mm is preferable); good contrast of black characters against a white background is required for clear photoprocess reproduction (glossies are not suitable for microfilm processing). A duplicate copy is required for indexing purposes.

A paragraph should appear at the end of the paper indicating the nature of the material and the means by which the interested reader may obtain copies directly. Use the following format:

Supplementary Material Available: Description of the material (no. of pages). Ordering information is given on any current masthead page.

V. Functions of Reviewers

The editors request the scientific advice of reviewers who are active in the area of research covered by the manuscript. The reviewers act only in an advisory capacity, and the final decision concerning a manuscript is the responsibility of the editors. The reviewers are asked to comment not only on the scientific content but also on the manuscript's suitability for *The Journal of Physical Chemistry*.

With respect to *Letters*, the reviewers are asked to comment specifically on the urgency of publication. **Authors are encouraged to suggest, when submitting a manuscript, names of scientists who could give a competent and objective evaluation of the work.** All reviews are anonymous and the reviewing process is most effective if reviewers do not reveal their identities to the authors. An exception arises in connection with a manuscript submitted for publication in the form of a comment on the work of another author. Under such circumstances the first author will, in general, be allowed to review the communication and to write a rebuttal. The rebuttal and the original communication may be published together in the same issue of the journal.

VI. Revised Manuscripts

A manuscript sent back to an author for revision should be returned to the editor within 6 months; otherwise it will be considered withdrawn. Revised manuscripts returned to the editor must be submitted in triplicate and all changes should be made by typewriter. **Unless the changes are very minor, all pages affected by revision must be retyped.** Revised manuscripts are generally sent back to the original reviewers, who are asked to comment on the revisions. If only minor revisions are involved, the editors examine the revised manuscript in light of the recommendations of the reviewers without seeking further opinions. If revisions are so extensive that a new typescript of the manuscript is necessary, it is requested that a copy of the original manuscript be submitted along with the revised one. In any case, a letter from the author should accompany the revised manuscript in which a detailed account of how the author has responded to the reviewer's comments is given. The dates of receipt of the original and the revised manuscripts will both appear in publication.

VII. Proofs and Reprints

Correspondence regarding accepted manuscripts should be directed to Journals Department, American Chemical Society, PO Box 3330, Columbus, OH 43210.

Galley proofs, original manuscript, cut copy, and reprint order form are sent by the printer directly to the author who submitted the manuscript. Authors of *Letters* are asked to enter any changes in their galley *by phone or Telex* within 72 hours of receiving the proofs; otherwise the proofreading will be done by the Journals Department.

The attention of the authors is directed to the instructions which accompany the proof, especially the requirement that all corrections, revisions, and additions be entered on the proof and not on the manuscript. Proofs should be checked against the manuscript (in particular all tables, equations, and formulas, since this is not done by the editor) and returned as soon as possible. Substantial changes in an article after it has been set in type are made at the author's expense. The filled-out reprint form must be returned with the proof. Reprint shipments are 5–6 weeks after publication, and bills are issued subsequent to shipment. Requests for reprints should be addressed to the author concerned; photocopies or tear sheets may be purchased from Business Operations, ACS.

A page charge covers in part the cost of publication. **Payment is not a condition for publication.** Articles are accepted or rejected solely on the basis of scientific merit. The charge per journal page is $25 for *Articles* and *Comments*. There are no charges for *Letters* or *Invited Articles*.

Figure 14 (continued)

Journal of Geophysical Research

Information for Contributors

Original scientific contributions on the physics and chemistry of the earth, its environments, and the solar system will be considered for publication in the *Journal of Geophysical Research*. Every part of a paper submitted for publication should convey the author's findings precisely and immediately to the reader. The author is urged to have his paper reviewed critically by his colleagues for scientific accuracy and clarity of presentation.

Typescript. All parts of the paper must be typed double-or, preferably, triple-spaced on good quality white paper 8½ by 11 inches (21.5 x 28 cm) with 1-inch (2.5 cm) margins at top, bottom, and sides. Corrasable bond, which smudges easily, and tissue paper are not acceptable. Each page of the typescript should be numbered in the top right-hand corner.

Authors are expected to supply neat, clean copy and to use correct spelling, punctuation, grammar, and syntax. Spelling and hyphenation of compound words follow the unabridged *Webster's Third New International Dictionary.* The metric system must be used throughout; use of appropriate SI units is encouraged. Following recommended style and usage expedites processing typescripts and reduces the chance of error in the typesetting.

Because text footnotes are expensive to set and distracting to the reader, they should be incorporated into the text or eliminated completely.

The typescript should be arranged as follows: (1) title page including authors' names and affiliations, (2) abstract, (3) text, (4) reference list, (5) tables, and (6) figure legends.

Abstract. The abstract should in a single paragraph (150 words or less) state the nature of the investigation and summarize its important conclusions. Listing the contents in terms such as 'this paper describes' or 'the paper presents' should be avoided. Use of the passive voice often indicates that the author is merely describing the procedure rather than presenting conclusions. References should not be cited in the abstract.

The abstract should be suitable for separate publication in an abstract journal and adequate for indexing.

When a paper is accepted for publication, the author will be requested to prepare camera-ready copy of the abstract for inclusion in *Eos.* At the same time he will select appropriate index terms for the paper.

Mathematics. All characters available on a standard typewriter must be typewritten in equations as well as in text. The letter l and the numeral 1 and the letter O and the numeral 0, which are identical on most typewriters, should be identified throughout the paper to prevent errors in typesetting. Any symbols that must be drawn by hand should be identified by a note in the margin. Authors will be charged for all corrections in galley proof for material that was handwritten in the typescript. Allow ample space (3/4 inch above and below) around equations so that type can be marked for the printer.

Alignment of symbols must be unambiguous. Superscripts and subscripts should clearly be in superior or inferior position. Since subscripts and superscripts cannot be set to align vertically (i.e., the superscript directly over the subscript),

the subscript will be set to precede the superscript: b_i
Fraction bars should extend under the entire numerator.

Barred and accented characters that are available for machine typesetting may be used. Symbols that are not available and therefore must be avoided are bars or accents that extend over more than one character, double or triple dots, and double accents (e.g., a circumflex over a bar). Angle brackets $< >$ can be substituted for overbars to represent averages, and accents over characters can be eliminated by the use of such symbols as ', *, and # set as superscript.

If an accent or underscore has been used to designate special type face (e.g., boldface for vectors, script for transforms, sans serif for tensors), the type should be specified by a note in the margin.

If the argument of an exponential is complicated or lengthy, exp rather than e should be used. Fractional exponents should be used instead of radical signs. Awkward fractional composition can be avoided by the proper introduction of negative powers. In text, solidus fractions (1/x) should be used, and enough parentheses should be included to avoid ambiguity. According to the accepted convention, parentheses, brackets, and braces are written in the order $\{[()]\}$.

Displayed equations should be numbered consecutively throughout the paper; the number (in parentheses) should be to the right of the equation.

References. A complete and accurate reference list is of major importance. Omissions, discrepancies in the spelling of names, errors in titles, and incorrect dates make citation annoying, if not worthless, to the reader and cast doubts on the reliability of the author as well.

Only works cited in the text should be included in the reference list. References are cited in text by the last name of the author and the date: [*Jones*, 1970]. If the author's name is part of the sentence, only the date is bracketed. Personal communications and unpublished data or reports are not included in the reference list; they should be shown parenthetically in text: (F. S. Jones, unpublished data, 1970).

References are arranged alphabetically by the last name of authors. Multiple entries for a single author are arranged chronologically. Two or more publications by the same author in the same year are distinguished by a, b, c after the year.

For laboratory, company, or government reports, information should be included on where the report can be obtained. For Ph.D. and M.S. theses the institution granting the degree and its location should be given.

References to papers delivered at meetings should include title of paper, full name of meeting, sponsor, meeting site and date. Citations of papers presented at meetings have been complicated by the recent practice of collecting manuscripts from participants, reproducing the manuscripts by offset, and distributing the offset collections to people attending the meeting. Such collections should not be cited as published works.

For references to books the page numbers of material being cited should be given.

Figure 15   Instructions to Authors of the *Journal of Geophysical Research.*

Names of periodicals should be written out in full or abbreviated according to the system employed by the Chemical Abstracts Service. One-word titles should always be given in full: *Science.* It is permissible to give only the initial page number of a paper, but preferable to give the range of pages. Samples:

Brandt, J. C., *Introduction to the Solar Wind,* p. 150, W. H. Freeman, San Francisco, Calif., 1970.

Choate, R., Lunar soil bulk density as determined from Surveyor data and laboratory tests, *Tech. Rep. 32-1443,* p. 35, Jet Propul. Lab., Pasadena, Calif., 1969.

Craig, H., Abyssal carbon 13 in the South Pacific, *J. Geophys. Res., 75,* 691-695, 1970.

French, B. M., and N. M. Short (Eds.), *Shock Metamorphism of Natural Materials,* pp. 520-528, Mono, Baltimore, Md., 1968.

McDonald, F. B., IQSY observations of low-energy galactic and solar cosmic rays, in *Annals of the IQSY,* vol. 4, edited by A. C. Strickland, p. 187, MIT Press, Cambridge, Mass., 1969.

Paschman, G., G. Haerendel, R. G. Johnson, and R. D. Sharp, Correlated variations of intensity and mean energy of electrons in the night side auroral zone (abstract), *Eos Trans. AGU, 52,* 322, 1971.

Shoemaker, E. M., E. D. Jackson, and M. H. Hait, Surficial and bedrock stratigraphy of the Apollo 12 landing site, paper presented at the Annual Meeting, Geol. Soc. of Amer., Washington, D. C., Nov. 1, 1971.

Soderblom, L. A., The disturbance ages of regional units in the lunar maria, Ph.D. thesis, Calif. Inst. of Technol., Pasadena, 1970.

Tables. Tables should be typed as authors expect them to look in print. Every table should have a title. Column headings should be arranged so that their relation to the data is clear. Footnotes should be indicated by reference marks (*, †, ‡, §) or by lower case letters typed as superiors. Each table must be cited in text. Authors are urged to use AGU's microform depository for tables of supporting data.

Illustrations. One set of illustrations suitable for the engraver and two clear copies for the reviewers should accompany the typescript. Original drawings (in black drawing ink on Bristol board or tracing cloth) or sharply focused glossy prints should be supplied for engravings. Lettering preferably should be 00 or 000 Leroy or equivalent. When printed, the smallest lettering or symbol should be at least 1/16 inch (1 1/2 mm) high; the largest lettering should not exceed 1/8 inch (3 mm). The following limiting dimensions apply: page height, 9 5/8 inches (24.5 cm); page width, 6 3/4 inches (17.4 cm); column width, 3 1/4 inches (8.3 cm).

Figures not cluttered with information that could be placed in the legend look neater and are easier to read. All details on the figures should be checked carefully, because corrections in galley proof (e.g., misspelled words, incorrect values, omitted symbols) necessitate making new plates. The cost of the new engraving will be charged to the author.

Each figure must be cited in text and must have a figure legend. Figures will be placed in the order mentioned in text.

Color figures and foldouts will be printed if the editor and reviewers judge them to be necessary for the proper presentation of the material. Copy for color figures must be supplied as type C prints the same size as the printed figure. The author should retain an identical print for checking the color proofs. The additional costs for these special figures must be borne by the author. The AGU editorial office should be consulted for current cost and special instructions.

If figures that have been published under copyright are to be reproduced in JGR, the AGU office must have written permission from the first publisher and the original author. Copies of letters of permission to republish should accompany the typescript.

Microform publication. Authors are encouraged to submit papers that are as concise as possible. Supporting material, such as tables of data, additional graphs, lengthy mathematical derivations, and extended background discussions, may be published as microfiche supplements. (Photographs with a wide tonal range are not suitable.) Summaries of articles and key figures may be submitted for publication in the journal, the detailed article being published on microfiche. All material published on microfiche is included in the microform editions of JGR (and therefore archived in libraries) and is available to individuals on order. Contact the editors or the AGU office for more information on this service.

Page charge. Authors' institutions are requested to pay a page charge. This payment entitles them to 100 reprints without further charge; additional reprints may be ordered at nominal rates. The page charge helps support the rapid publication of JGR and prevents the necessity of stringently restricting the number of pages to be published each year. The charge to authors for publishing on microfiche is substantially less than the publication charge.

Submitting the paper. Four copies of the complete typescript and three sets of figures should be sent to the appropriate editor. Check the inside cover of a recent issue for the names and addresses of the current editors.

Figure 15 (continued)

CANADIAN
JOURNAL
OF
MICROBIOLOGY

INSTRUCTIONS
TO
AUTHORS

The Canadian Journal of Microbiology (Can. J. Microbiol.) publishes, in either English or French, **articles**, **notes**, and the occasional **review**. **Articles** are reports of research in any field of microbiology and must be original contributions to science. **Notes** may be brief reports of work that is largely confirmatory, advances in knowledge arising as by-products of broader studies, or descriptions of research techniques or developments in instrumentation. Notes should not be longer than four printed pages (about 12 manuscript pages). They should have an Abstract, but should not be divided into Introduction, Materials and methods, Results, and Discussion sections. **Reviews** are critical appraisals of the literature in areas of microbiology that are of interest to a broad spectrum of microbiologists. All contributions are subject to the normal reviewing process.

Publication is facilitated if authors check very carefully the symbols, abbreviations, and technical terms for accuracy, consistency, and readability and ensure that manuscripts and illustrations meet the requirements outlined below. It is helpful to examine the journal itself for details of layout, especially for tables and reference lists.

It is the policy of the Journal to allow 6 months for revision of papers which require further experimental work, and 3 months for papers which require only revision. Authors will be reminded once and if there is no response, the manuscript will be considered withdrawn.

Authors will receive galley proofs but the page proofs will be made up before the galley proofs are returned. The printers will correct the page proof **only for typographical errors** as indicated on the galley proofs returned by the authors. Authors are asked to take the greatest care with the preparation of their papers. The manuscript which is submitted or returned to the Editor after final revision will be the one that is printed. It will not be possible to make alterations in proof except for the correction of typographical errors.

Recommendations of the Nomenclature Committee of the International Union of Biochemistry on the Nomenclature and Classification of Enzymes (Academic Press, New York, 1979) should be followed. Abbreviations and contractions of the names of substances, procedures, etc., must be defined the first time they occur or in a footnote on the title page. Symbols and unusual or Greek characters should be identified clearly; superscripts and subscripts should be legible and carefully placed, and they should be explained by marginal notes when necessary.

Authors who describe experiments on animals are required to give assurance that the animals were cared for in accordance with the principles of the *Guide to the Care and Use of Experimental Animals*, Vol. I. This guide is available from the Canadian Council of Animal Care, 1105–151 Slater St., Ottawa, Ont. K1P 5H3, $3/copy (soft cover, 112 pages).

Title—The first page of the manuscript should have only the title, the authors' names, the authors' affiliation, and any necessary footnotes. The authors' address must be the institution where the work was done. Authors' present address, if different, should be a footnote. It is essential that the title be clear, concise, and informative of the contents of the paper. Running titles are not published. It is desirable to indicate in a footnote the author to whom reprint requests should be addressed.

Abstract—An abstract of not more than 200 words, typed on a separate page, is required for each article or note. Review articles should have a table of contents instead of an abstract. No abbreviations should appear in the abstract.

References—Each reference should be denoted in the text by the author and date in parentheses as shown below. If two or more publications are listed for the author or authors in the same year they

THE MANUSCRIPT

Desiderata—Scientific merit and originality are the two most desirable qualities of any paper. In addition, papers must be clearly and concisely presented and suitable for a readership interested in microbiology. Methods, figures, and footnotes to tables should be so written that others can repeat or expand the work being reported. The body of the paper should address itself to the problems outlined in the introduction. The discussion should be relevant and the conclusions drawn adequately supported by the data.

General—All parts of the manuscript, including footnotes, tables, and captions for illustrations, should be typewritten, **double-spaced**, on one side only of white paper 21.5 by 28 cm, with margins of 4 cm. Do not underline unless the material is to be set in italics. Use capital letters only when the letters or words should appear in capitals in the printed paper. Indent the first line of all paragraphs in the text and of all captions and footnotes. **The original manuscript and two duplicate copies are required.** Double-sided copies are not acceptable. Each page of the manuscript *must* be numbered and it is helpful to reviewers if each line on each page is also numbered. Tables and captions for illustrations should be on separate pages and placed after the text.

Spelling should follow that of *Webster's Third New International Dictionary* or the *Oxford English Dictionary*. Authors are responsible for consistency in spelling. **Abbreviations, nomenclature, and symbols for units of measurements** should conform to international recommendations. **Metric units** should be used or metric equivalents given and the use of SI units (Système international d'unités) is encouraged. This system is explained in the *Metric Practice Guide* (1979) published by the Canadian Standards Association (178 Rexdale Blvd., Rexdale, Ont., Canada M9W 1R3) and in *Quantities, Units and Symbols* (1971) published by the Symbols Committee of the Royal Society (6 Carlton House Terrace, London, England SW1Y 5AG). As a general guide for biological terms the *Council of Biology Editors Style Manual* (American Institute of Biological Sciences) is recommended. For enzyme nomenclature *Enzyme Nomenclature 1978*:

are differentiated by a, b, c, etc., placed after the year, without space.
One author: (Smith 1977), (Rogers 1969, 1979), (Jones 1977; Smith 1965), (Brown 1965a, 1965b).
Two authors: (Smith and Rogers 1976).
Three or more authors: (Smith et al. 1975).
If the names of the authors form part of the text, only the date should appear in parentheses, as in "Miller (1975) reported that...". The reference list should be placed at the end of the text and the references should be listed in alphabetical order in the form used in current numbers of the journal. In references to papers in periodicals, titles and inclusive page numbers are required. The names of serials are abbreviated in the form given in *CASSI (Chemical Abstracts Service Source Index*, Chemical Abstracts, P.O. Box 3012, Columbus, OH, U.S.A. 43210). For serials not given in the guide, the abbreviated name is constructed from the *NCPTWA Word-Abbreviation List*, 1971 Edition, American National Standards Institute, Standards Committee Z39, and its supplements. Papers submitted, work from unreviewed publications (e.g., abstracts, proceedings of meetings), or theses are not allowed in the reference list but should be quoted within parentheses in the text. Papers "in press" may be listed among the references. Authors must give assurances to the Editors that the paper has been accepted. Volume and page numbers must be completed at the galley proof stage; if the paper has not appeared at that stage, the Journal will list such papers as footnotes.

Footnotes—Footnotes to material in the text should not be used unless they are unavoidable, but their use is encouraged in Tables. Where used in the text, footnotes should be designated by superscript arabic numerals in serial order throughout the manuscript except in Tables. Each footnote should be placed at the bottom of the manuscript page where reference to it is made.

Writing numbers—In long numbers the digits should be separated into groups of three, counted from the decimal marker to the left and right. The separator should be a space and not a comma, period, or any other mark, e.g. 23 562 987 and not 23,562.987. The decimal marker should be a point, e.g., 0.1 mL and not 0,1 mL. The sign '×' should be used to indicate multiplication, e.g., 3×10^6 and not 3.10^6.

Figure 16 Instructions to Authors of the *Canadian Journal of Microbiology.*

Equations—These must be set up clearly in type, triple-spaced. They should be identified by numerals in square brackets placed flush with the left margin. In numbering, no distinction is made between mathematical and chemical equations. Routine **structural formulae** can be typeset and need not be submitted as figures for direct reproduction, but they must be clearly depicted.

Tables—Tables should be numbered with arabic numerals, have a brief title, and be referred to in the text. Column headings and descriptive matter in tables should be brief. Vertical rules should not be used. A copy of the journal should be consulted to see how tables are set up and where the lines in them are placed. Footnotes in tables should be designated by symbols or superscript small italic letters. Descriptive material not designated by a footnote may be placed under a table as a **Note**.

Supplementary material—The National Research Council of Canada maintains a depository in which supplementary material such as extensive tables or data, detailed calculations, and colored illustrations may be placed. Authors wishing to use it should submit their complete work for examination by the editors and mark the part to be considered for deposition. For some papers, the editors may suggest that portions be placed in the depository. Photocopies of material in the depository may be obtained, at a nominal charge, from: Depository of Unpublished Data, CISTI, National Research Council of Canada, Ottawa, Canada K1A 0S2.

Permission to reprint—Whenever a manuscript contains material (tables, figures, charts, etc.) that is protected by copyright, it is the obligation of the author to secure written permission from the holder of the copyright.

Nomenclature of microorganisms—Authors are reminded that there was a revision of the *International Code of Nomenclature of Bacteria* effective 1 January 1976. A new name is not validly published until a note containing the name is also published in the *International Journal of Systematic Bacteriology*. Microorganisms and viruses should be given strain designations consisting of letters (usually two) followed by serial numbers. It is generally advisable to use the worker's initials or a descriptive symbol of locale or laboratory. Each new isolate will then be given a new (serial) designation (AB1, AB2, etc.). Genotypic and phenotypic symbols should not be included.

information about the symbols in current use, consult reviews by Bachmann and Low (Microbiol. Rev. 44: 1–56, 1980) for *Escherichia coli* K-12; Sanderson and Hartman (Microbiol. Rev. 42: 471–519, 1978) for *Salmonella typhimurium*; Holloway et al. (Microbiol. Rev. 43: 73–102, 1979) for *Pseudomonas*; and Henner and Hoch (Microbiol. Rev. 44: 57–82, 1980) for *Bacillus subtilis*.

Viruses—In the genetic nomenclature of bacterial viruses (bacteriophages), no distinctions are made between genotype and phenotype. Genetic symbols may be one, two, or three letters.

Transposable elements and plasmids—Nomenclature of transposable elements (transposons, Mu) should follow Campbell et al. (Gene, 5: 197–206, 1979), and for plamids, should follow Novick et al. (Bacteriol. Rev. 40: 168–189, 1976).

ILLUSTRATIONS

General—Each figure, or group of them, should be planned to fit into the area of either one or two columns of text. The maximum finished size of a one-column illustration is 7.8 × 21 cm and a two-column illustration is 16 × 21 cm. The figures (including halftones) are numbered consecutively in arabic numerals, and each one must be referred to in the text but should be self-explanatory. All terms, abbreviations, and symbols must correspond with those in the text. Only essential labeling should be used, with detailed information given in the caption. Each illustration should be identified by the authors' names and title of paper, preferably written below the illustration, at the left.

Line drawings—The original drawings or one set of glossy photographs and two sets of clear copies are required. Particular care should be taken that prints be well focused. Line drawings as submitted should not exceed 20 × 28 cm; larger illustrations are unacceptable. Drawings should be made with India ink on plain or blue-lined white paper or other suitable material. Any coordinate lines to appear should be ruled in. All lines must be sufficiently thick to reproduce well, and decimal points, periods, dots, etc., must be large enough to allow for any necessary reduction. Letters and numerals should be made neatly with a printing device (**not a typewriter**) or come from sheets of printed characters and be of such size that the

Genetic nomenclature

Bacteria—The genetic properties of bacteria are described in terms of genotypes and phenotypes. The phenotypic designation describes the observable properties of an organism. The genotype refers to the genetic constitution, usually in reference to a standard wild type. Use the recommendations of Demerec et al. (Genetics, 54: 61–74, 1966) as a guide in employing these terms. (*i*) Phenotypic designations must be used when mutant loci have not been identified or mapped. Phenotypic designations generally consist of three-letter symbols, not italicized, with the first letter capitalized. Superscript letters may be used (e.g., Str^r for streptomycin sensitivity). (*ii*) Genotypic designations are composed of three-letter locus symbols written in lowercase italics. Wild-type alleles are indicated by positive superscripts (e.g., his^+). If several loci control related functions, they are distinguished by italicized capital letters following the locus symbols (e.g., *hisA*, *hisB*). Mutation sites are indicated by putting the serial isolation numbers (allele numbers) after the locus symbol. (*iii*) Where large numbers of strains are used, it may be useful to show a table of strains. Deviations from normal use should be defined. For more detailed

smallest character will not be less than 1 mm high when reduced. The same type of lettering should be used for all figures in any one paper. Care should be taken to have the drawing and lettering in good proportion so that both can take the same reduction.

Photographs—**Three sets of all photographs are required:** (1) one set mounted on **light** cardboard, ready for reproduction; (2) two more sets, equally good, but preferably unmounted. Prints must be of high quality, made on glossy paper, with strong contrasts. The copies of reproduction should be **trimmed to show only essential features and** mounted on white cardboard, with **no space** between those arranged in groups. A photograph, or group of them, should be planned to fit into the area of either one or two columns of text **with no further reduction.** Electron micrographs or photomicrographs should include a scale line or bar directly on the print.

Color illustrations—Illustrations may be accepted for reproduction in color subject to the Editor's decision that the use of color is essential. Authors will be responsible for all costs and must accept other conditions which may be obtained from the Publishing Department.

Figure 16 (continued)

Revised January 1983

Chapter 9

Post-Printing Activities

In well-managed journals, the post-printing activities, as well as the copy editing and printing, should be the responsibilities of managing editors or business managers. The scientific editors will have certain concerns and interests as indicated below; however, as with copy editing and printing, their active participation should not be expected to go much beyond the provision of advice or feedback.

Distribution and Marketing

Scientific editors should be concerned that their journals have as large and broad a distribution as possible because authors are generally more attracted to journals with large circulations. Thus, editors should encourage their publishers in any efforts to increase circulation of their journals. That encouragement could take the form of reviewing existing distribution lists to identify any obvious gaps, of promoting the journal among scientific colleagues at meetings and conferences, and of providing feedback from existing subscribers. On the latter point, the scientific editor is the main contact point between the journal and the community it serves and will therefore receive any complaints. Any adverse experience reported by subscribers should be acknowledged by the editor and referred immediately to the managing editor or business manager for follow-up action. It might be that complaints about issues being damaged in the mails could lead to a review of packaging methods or that reports of missing issues reveal shortcomings in the registration of new addresses. Unless such complaints are scrupulously acknowledged and relayed by editors, they cannot be followed up for corrective action. Not even a single complaint can be ignored because it may be symptomatic of a problem that affects a number of subscribers who did not complain but may simply cancel their subscriptions.

128

The marketing of journals is a skilled, professional activity that is best left in the hands of experts who know the relative values of different promotion techniques. However, those experts, if they are smart, will call upon scientific editors for assistance in the definition of markets, and in the preparation or review of promotional brochures. For example, editors and their associate editors, may be able to supplement mailing lists with names of colleagues or of scientists in foreign countries. A brochure for a promotional campaign might highlight a new direction or thrust in a journal's editorial scope as an added inducement to new subscribers. Scientific editors can contribute valuably to the preparation of promotional material for their journals and should be consulted in the formulation of any promotional campaigns.

Secondary Services

A large part of any journal's visibility lies in its coverage by the important current-awareness vehicles, abstracts journals, or indexing services in its field. Publications such as *Chemical Abstracts*, *Biological Abstracts*, *Index Medicus*, *Current Contents*, and *Science Citation Index* are important services, much consulted by scientists for current awareness or for comprehensive literature searches. Journals not covered by these secondary services tend to be ignored, or "lost." All of the various current-awareness, abstracts, or indexing services have their own criteria for the inclusion of journals in their coverage; some strive for comprehensiveness, others are more selective. In any case, scientific editors should try to see that their journals are covered by every such information service that is significant and appropriate. This may mean calling the attention of the secondary services to the journal by sending a sample issue and a well-reasoned letter that sets forth and substantiates the importance of the journal in its field. Of course, major journals do not have this problem; they cannot be ignored by any information service. Nevertheless, the managers and editors of all journals should realize that coverage by information services represents the best possible publicity and should therefore cooperate with those services to the greatest extent possible. At the minimum, this cooperation should include the provision of free subscriptions to all secondary services, delivered as quickly as possible; some journals go so far as to supply advance copies or page proof. Garfield has written extensively about what publishers and editors can do to make it easier for their journals to be covered by *Current Contents* and *Science Citation Index*.[10] His main concerns are format of the contents pages, quality of printing, placement and identification of authors' addresses, and treatment of cited references. The contents page should be easy to scan, and this means having a clear distinction between titles and authors, by the use either of a two column format or of different type faces. Poor print quality, either too heavy and smeared or too light, will not be legible on photoreproduction and requires recomposition. If type size is either too small or too large, photoreproduction to fit the *Current Contents* page size is a problem. Authors' addresses should appear directly under the names at the head of articles; when there is more than one address, the authors' affiliations should be clearly identified. Cited references should be assembled in a reference list at the end of each article and should not be given as footnotes on the page where they are mentioned in the text. The importance of *Current Contents* and *Science Citation Index* as secondary services would alone justify serious consideration of these recommendations which would also, however, make journals easier to scan, to index, and to enter into data bases. Journals that adopt these sound recommendations will therefore participate more effectively in scientific communication and information retrieval.

Information Retrieval

If a scientific paper is to contribute to the advancement of knowledge, it must be accessible to the scientific community and it should be readily retrievable. This accounts for the overwhelming preference of scientists to publish in the periodic journals rather than in books, monographs, or isolated collections of papers such as conference proceedings. A paper published in a regular issue of a periodic journal can readily be tracked down from any one of the following: authors' names (from the author index), title (from the contents pages or subject index), and page number (with the volume number or year). This may be contrasted with a search for a paper published in a conference proceedings where the main problem is tracking down the book, particularly if the searcher has forgotten the names of the editors and maybe even the title. Thus, the first contribution that a journal can make to information retrieval is already in place, i.e., the journal's identity, its name. Of course, the name should appear prominently on the front cover of each issue and also on the spine to help in identification of unbound issues. A change in the name of a journal results in loss of identity and should be avoided as far as possible. If a change is deemed essential, it should be made at the start of a new volume and with ample advance notice to all secondary services so they can link the new with the old.

Another important part of a journal's identity is the numeric or alphabetic code that is used by information services to uniquely identify each serial publication. There are two such codes in use at the present time: one is the International Standard Serial Number (ISSN) which is an eight-digit code; the other is CODEN, a five or six uppercase letter code based upon the journal's title. ISSN codes are assigned either by a national center of the International Serials Data System or by its international center at 20 rue Bachaumont, 75002 Paris, France. CODEN codes are assigned by the American Society for Testing and Materials through the International CODEN Service of Chemical Abstracts, Ohio State University, Columbus, Ohio 43210. Both of these machine readable codes should be displayed on an upper corner of the front cover of each issue of the journal.

Scientific editors can make their journals more useful for information retrieval by paying special attention to titles, key words, abstracts, authors' names, and cited references. For information retrieval purposes, a title is possibly the most important part of a paper, but it may also be the part that receives the least thought by the authors. Titles appear in all current awareness services, in computer printouts from information retrieval systems, in subject indexes, and often in citations. It is therefore important that titles describe the contents of papers as fully and specifically as possible. For example, a title such as "Studies on Pine Needles. Part 10," says very little. What kind of studies are being reported — chemical, physiological, morphological? Better would be "Studies on Pine Needles. Part 10: Analysis of Terpenes"; even better is "Studies on Pine Needles. Part 10: Analysis and Identification of Terpene Components"; however, the best would be "Identification and Quantitative Analysis of the Terpene Components of Pine Needles," thus eliminating the meaningless word "Studies" and the unnecessary "Part 10."

The producers of data bases request journals to list key words for each paper. These are words that generally cannot be fitted into a title but that help to further identify the subject matter of the paper. Authors can be asked to provide key words which subsequently should be scrutinized by the editor to see that they are both appropriate and complete. When key words are given, they are usually placed immediately after the abstract in the printed paper.

Abstracts are intended to assist readers of journals by giving brief summaries of the results and conclusions reported in the papers. Because most abstracting services now use authors' abstracts directly, editors should stress the need for careful preparation and should check that each abstract is truly informative.

Information retrieval from author indexes is greatly aided if journals require authors to give their first names in full rather than use just initials. There may be a considerable number of scientists who are designated as J. A. Jones; the number will be decreased if Dr. Jones uses Jonathan A. Jones and may be reduced to one if he uses Jonathan Alonzo Jones. It is in everyone's best interests for journals to cooperate in the attempt to reduce the number of homographs among authors' names.

Finally, information retrieval is facilitated if journals require that references be complete, *including first and last page numbers and titles of the cited papers.* The inclusive page numbers show how long the cited paper is, and the titles let a reader pick out pertinent papers from the reference list without searching back through the article to see in what context they were cited. The adoption of a standard form of citations by all journals is highly desirable and should be possible. However, some parts of the scientific community seem bound by tradition or conservatism and are reluctant to change. Probably the best that can be done is to urge all editors to adopt the complete form of citation, as described above, whenever possible.

References

1. **Broad, W. J.** 1980. Would-be academician pirates papers. Science **208:**1438–1440.
2. **Broad, W. J.** 1981. The publishing game: getting more for less. Science **211:**1137–1139.
3. **Christiansen, D.** 1983. Spectral lines. Multiple exposure. IEEE Spectrum **20:**31.
4. **Cupas, C., P. von R. Schleyer, and D. J. Trecker.** 1965. Congressane. J. Am. Chem. Soc. **87:**917–918.
5. **Day, R. A.** 1975. How to write a scientific paper. ASM News **41:**486–494.
6. **Day, R. A.** 1983. How to Write and Publish a Scientific Paper, 2nd ed. ISI Press, Philadelphia.
7. **Editorial Board.** 1961. Editorial announcement. J. Infect. Dis., Vol. 108, no. 3, May–June.
8. **Editorial Board.** 1961. An unfortunate event. Science **134:**945.
9. **Editorial Board.** 1965. Editorial note. Can. J. Phys. **43:**2383.
10. **Garfield, E.** 1980. In recognition of journals which prove that change is possible, p. 482–487. *In* Essays of an Information Scientist, Vol. 3. ISI Press, Philadelphia.
11. **Garfield, E.** 1982. More on the ethics of scientific publication: abuses of authorship attribution and citation amnesia undermine the reward system of science. Current Contents, no. 30, 26 July, pp. 5–10.
12. **Goudsmit, S. A.** 1970. Editorial. Phys. Rev. Lett. **25:**419–420.
13. **Greenstein, J. S.** 1965. Studies on a new, peerless contraceptive agent: a preliminary final report. Can. Med. Assoc. J. **93:**1351–1355.
14. **Harnard, S.** 1982. Peer commentary on peer review. The Behavioral and Brain Sciences **5:**185–186.
15. **Harrison, L. G., J. A. Morrison, and R. Rudham.** 1958. Chloride ion diffusion in sodium chloride. Trans. Faraday Soc. **54:**106–115.
16. **Honig, W.** 1981. Editorial. Speculations in science and technology **4:**1.
17. **Houghton, B.** 1975. Scientific Periodicals: Their Historical Development, Characteristics and Control. Linnet Books, Hamden, CT.
18. **Kline, N. S., and J. C. Saunders.** 1959. Psychochemical symbolism. J. Psychol. **48:**279–283.

19. **Kronick, D. A.** 1976. A History of Scientific and Technical Periodicals. Scarecrow, Metuchen, NJ.

20. **Lindsay, D., J. A. Howard, E. C. Horswill, L. Iton, K. U. Ingold, T. Cobbley, and A. Ll.** 1972. The bimolecular self-reactions of secondary peroxy radicals. Product studies. Can J. Chem. **51:**870–880.

21. **Parkinson, C. N.** 1962. In-laws and Outlaws, pp. 166–167. Greenwood Press, Westport, CT.

22. **Price, D. J. de S.** 1961. Science since Babylon. Yale University Press, New Haven, CT.

23. **Price, D. J. de S.** 1964. Ethics of scientific publication. Science **144:**655–657.

24. **Shishido, A.** 1960. Letter to the editor. Nature **286:**437.

25. **Smith, A. S.** 1955. The stereochemistry of octahydrohexairon: a molecular raft. Chem. and Ind., pp. 353–354.

26. **Weber, R. L.** 1982. More Random Walks in Science. Institute of Physics, U.S. dist., Heyden, Philadelphia. From an editorial review, 1983. Physics Today, p. 77.

27. **Ziman, J. M.** 1966. Public Knowledge: The Social Dimension of Science. Cambridge University Press, Cambridge, U.K.

Index